Galileo

Decisive Innovator

GALILEO

DECISIVE INNOVATOR

MICHAEL SHARRATT

CAMBRIDGE
UNIVERSITY PRESS

PUBLISHED BY THE PRESS SYNDICATE OF THE UNIVERSITY OF CAMBRIDGE
The Pitt Building, Trumpington Street, Cambridge, United Kingdom

CAMBRIDGE UNIVERSITY PRESS
The Edinburgh Building, Cambridge CB2 2RU, UK http://www.cup.cam.ac.uk
40 West 20th Street, New York, NY 10011-4211, USA http://www.cup.org
10 Stamford Road, Oakleigh, Melbourne 3166, Australia

First published 1994 by Blackwell Publishers Oxford
Reissued by Cambridge University Press 1996
Reprinted 1999

Printed in the United Kingdom at the University Press, Cambridge

Typeface Palatino *System* QuarkXpress®

A catalogue record for this book is available from the British Library

ISBN 0 521 56219 8 hardback
ISBN 0 521 56671 1 paperback

For Ted Sharratt

Contents

Illustrations

General Editor's Preface

Our society depends upon science, and yet to many of us what scientists do is a mystery. The sciences are not just collections of facts but are ordered by theory, and this is where Einstein's famous phrase about science being a free creation of the human mind comes in. Science is a fully human activity; the personalities of those who practise it are important in its progress and often interesting to us. Looking at the lives of scientists is a way of bringing science to life.

Galileo, the last man of science whom we customarily call by his first name, was the first scientific star: his work brought him European fame, and in his own day and since he stood out from even the ablest of his predecessors and contemporaries. If we had to name one founder of modern physical science, with its dependence upon mathematical reasoning in defiance of common sense, then most of us would pitch upon Galileo. He insisted that the same physics must apply to the Earth and the heavens; and he devised, and sometimes carried out, experimental tests of his reasoning. He did not invent the telescope but he perceived its importance, enabling us to see things (however blurred in the early models) never seen before, with momentous consequences for accepted beliefs about the world. Denounced to the Inquisition, he was, in his old age, condemned to house arrest after a show trial, which made him a kind of martyr to science and has assured his continuing fame.

He does not fit our picture of the physicist, doing his most brilliant work at the age of about twenty. He was no youthful prodigy, and was propelled into the limelight only by his work on the telescope done in his forties. His fundamental challenge to the physics of Aristotle came later, and by the time he faced the Inquisition he was nearly seventy. His most important scientific treatise was written later; after the trial, he recovered his zest for science, and never lost his intellectual liveliness.

He was no austere and cloistered academic. He wrote brilliant invective, never suffering fools gladly. His self-confidence was enormous, and he tended to regard astronomy and mechanics as the territory on which he had staked out a claim, and on which trespassers risked being prosecuted. He was in a hurry; he knew that he was right about the Earth going round the Sun and the new inertial physics that made sense of this idea, and he refused to say that this Copernican theory was a mere calculating device. A milder man might have got the new ideas admitted gradually (probably not in his lifetime); but Galileo's temperament put him on collision courses. He was, in his own estimation at least, a good Catholic; and he felt urgently that the Church must not commit itself to obsolete science. Belatedly, his Church has now vindicated him.

Michael Sharratt succeeds in bringing Galileo to life. His researches on the science, the theology and the general history of the early seventeenth century make him an ideal biographer for this brilliant and pugnacious man, who saw further than most contemporaries. He transports us into the controversies in which Galileo was engaged, on serious issues that are still with us in various forms, and also illuminates Galileo's Venice, Florence and Rome. It is an excellent book.

David Knight
University of Durham

Acknowledgements

My first thanks go to my colleagues and students at Ushaw College for their interest and support while this book was in the making; to the governors of the College for granting me two sabbatical terms for research; to Jan Rhodes for many things, but specifically for looking after the College's library and archives while I was on sabbatical leave and for making available the illustrations for this book when she succeeded me as librarian in January 1993. Warm thanks, next, to the Master and community of Campion Hall, Oxford, who made me very welcome from January to July of 1991 and provided unobtrusive but constant support throughout that time and afterwards. I am very grateful to Alistair Crombie, John Roche and Beth Rainey, Keeper of Rare Books in the University of Durham, who were all generous with their time, diverse expert knowledge and encouragement at an early stage of my research. I owe special thanks to my brother Barney for taking the photographs which illustrate this book, to David Knight, the general editor of this series, for his kindly and encouraging help at all stages, and to the publisher's patient editorial assistants. This list would be incomplete without the names of Rosemary Bayne, Susan Biggin, Brian Ferme, Peter Fleetwood and Ted Sharratt. My greatest debt, as anyone familiar with Galileo will know and as will be obvious throughout the book, is to Antonio Favaro, the editor of the splendid *Edizione Nazionale* of Galileo's works.

Michael Sharratt
Ushaw College

1

The Strangest Piece of News

Sir Henry Wotton, the English Ambassador to the Republic of Venice, was taking a risk when he wrote to the Earl of Salisbury, sending King James I what he called 'the strangest piece of news' that the King had 'ever yet received from any part of the world'. The news came in the Latin booklet he enclosed, entitled *The Sidereal Message* (*Sidereus Nuncius*) (3: 53–96. The reference is to volume 3, pages 53–96 of the *Edizione Nazionale* of Galileo's works).[1] He had not wasted any time: the work, written by the mathematical professor at Venice's University of Padua, had appeared that very day, 13 March 1610. Though Wotton did not name the author, Galileo Galilei, he described the sensational news briefly and accurately. The professor had used an optical instrument, which enlarges objects and brings them nearer; the instrument was invented in Flanders but improved by the professor. With his instrument he had 'discovered four new planets rolling about the sphere of Jupiter, besides many other unknown fixed stars'. He had settled the long disputed nature of the Milky Way and found that the Moon is not perfectly spherical 'but endued with many prominences' and, strangest of all to Wotton, illuminated with the Sun's light reflected from the Earth: Wotton was not sure he had got this last point right, or perhaps not sure that it was credible, for he adds 'as he seemeth to say'. But the upshot was that all astronomy and astrology had been overturned – according to Wotton, rather than the booklet itself – and naturally all corners of Venice were full of the news. Before concluding by promising to send one of the new instruments by the next ship, Wotton realized that in the next post he might have to admit that he had been a credulous fool, so he covered himself neatly against this risk by saying: 'And the author runneth a fortune to be either exceeding famous or exceeding ridiculous.'[2]

We are so accustomed to Galileo's exceeding fame that it is useful to remember that it was his telescopic discoveries that made him famous. If he had died a year earlier at the age of forty-five, he would now be known only to historians of science. He would not be completely forgotten: in fact, much of the work on which his enduring fame rests was already well advanced before he ever heard of the new optical instrument, but it had not got beyond rough working papers which were known only to a few friends. He did have some reputation in Italy and abroad for his well-designed and handsomely crafted calculating instrument, the geometrical compass, with its printed manual of instructions. His position was an honourable one, mathematical professor at the renowned University of Padua, and he was becoming useful to the rulers of his native Tuscany. Even if he had died prematurely there were pupils and friends who would have kept his memory alive for a while. Sooner or later scholars would have come across him and, if the manuscripts of such a person had survived, perhaps a local society would have been founded to reconstruct his discoveries. But he would never have been famous.

His reputation would be very much like that of his contemporary, the Englishman Thomas Harriot (1560–1621), who had observed the Moon through a telescope and made a drawing of it on 5 August 1609, when Galileo had not got beyond showing how much could be seen on land and sea with his telescope.[3] Harriot was a versatile genius: English people can compare him with Galileo without fear of being accused of chauvinism. Nor did he keep his discoveries entirely to himself: he had his friend, Sir William Lower, observing the Moon through all its phases just at the time Galileo was observing the newly discovered satellites. Lower was able to see in the new Moon the earthshine that Wotton was to find so strange and he thought the full Moon looked 'like a tarte that my Cooke made me the last Weeke – here a vaine of bright stuff, and there of darke, and so confusedlie all over'.[4] But, though Harriot was a major innovator in mathematics and physics, he did not publish his findings, whereas Galileo did.

Not only did Galileo publish his telescopic discoveries, he did so before anyone else produced anything on the subject and he followed them up with more; from this time on he displayed the rare gift of popularizing his views in language that any educated person who was not completely set against new ideas could enjoy and more or less

1 Galileo: the frontispiece of Viviani's *De locis solidis* (Florence, 1701).

understand. He also had a very good conceit of his own amazing abilities and was to find it difficult to allow others a share in the glory that he thought was his alone. He is most widely remembered for his brilliant but unsuccessful campaign to use the novel telescopic discoveries to gain acceptance for the (more or less) Sun-centred system of the universe proposed decades earlier by Nicolaus Copernicus (1473–1543)

— 3 —

in his *De Revolutionibus Orbium Coelestium* of 1543. The first set-back to Galileo's campaign came in 1616, only a few years after he discovered the planets rolling about Jupiter: the Congregation of the Index ruled that Copernicanism could be treated by Catholics only as a calculating device. For Galileo himself, though not for his cause, the definitive defeat came in 1633 when he was condemned by the Roman Inquisition for his flawed but fascinating *Dialogue on the Two Chief World Systems (Dialogo sopra i due massimi sistemi del mondo)*, which compared the geocentric approach of Ptolemy with the heliocentric system of Copernicus. That notorious condemnation has made Galileo into an enduring symbol of scientific freedom and an embarrassing famous son to his own Church. It is remarkable that in his declining years, despite bereavement, approaching blindness and humiliating restrictions on his freedom of movement and association, he was able to draw together and complete his earlier pioneering studies, particularly on motion, in his greatest work, the *Discourses on Two New Sciences (Discorsi e dimostrazioni matematiche intorno a due nuove scienze)* of 1638.

Even the brief sketch in the preceding paragraph is sufficient to raise the question: how should Galileo's life be presented to modern readers who perhaps know little of him beyond the notion that he is a founder, perhaps the founder, of modern science? Stillman Drake, whose publications have done a great deal over recent decades to draw attention to how Galileo actually worked, has no doubt: 'The most faithful portrait of Galileo as a scientist is one that shows him in the role of the pioneer modern physicist, and not in that of an over-zealous Copernican astronomer.'[5] There is a good deal to be said for Drake's view. Much of Galileo's enduring reputation rests on his having pioneered ideas which have been enormously fruitful in physics. Galileo himself knew that he was starting something new. But the fact remains that he deliberately devoted a large part of two whole decades to furthering Copernicanism, decades in which, apart from ill health, he was at the height of his intellectual powers. During those decades he did not respond to the occasional urgings.of a few knowledgeable friends to complete his studies of how bodies move on or near the Earth. This may be a matter for regret, but it is what happened. It is, of course, important not to read the first forty-five years of his life as though they had been leading up to the public espousal of Copernicanism in 1610. Galileo *was* looking to escape from the daily tedium of humdrum teaching but he was no Churchill, fretting away his powers until the moment of destiny should arrive; in any case, astronomy had never occupied the chief place in his studies. Drake is also right to insist that Galileo's work in physics is of greater significance for later science than anything, however impressive, which he contributed to astronomy.

Yet his campaign to establish Copernicanism included a major contribution to physics, namely the unseating of Aristotelian philosophers, who hitherto had assumed they were the final court of appeal on what motion is. An essential step was to reduce them to mere members of a divided jury and to show them that their case against a moving Earth was not proven. This Galileo was to do in the most important section of his *Dialogue* of 1632, so there need be no distortion or downgrading of physics in following Galileo's life as it comes: samples of his work in physics can still be introduced in their proper place and their lasting significance can be indicated. By the same token, this brief biography will give a good deal of space not just to Galileo's astronomy and physics but also to his ways of dealing with philosophical and theological questions. It may seem to some that such topics are not strictly relevant to the history of science. That seems to me an artificial compartmentalization, but in any case they certainly seemed relevant to Galileo. The man who wanted to establish a new approach to natural philosophy and a new system of the world could not avoid such questions, even if, as seems clear, he knew that the crucial area was physics, especially the physics of motion.

One good way to introduce Galileo's treatment of all these topics to modern readers is to make the most of the clear, forceful and exciting expositions which he wrote for the educated public of his time. There is a great deal of controversy about his personality, originality, scientific judgement and philosophical ideas. The vast scholarly literature about him contains few topics of importance that are undisputed in their interpretation, but it is generally acknowledged that he was a great writer and teacher – though even as a teacher he is often accused of oversimplification and sometimes of dishonesty. The sudden fame that followed *The Sidereal Message* provided Galileo with the platform to display his remarkable gifts. He seized this heaven-sent opportunity to reshape his life. Certainly there is much more to him than a zealous Copernican astronomer – whether he was overzealous will appear in the following chapters. But it was Galileo himself who chose to concentrate his powers on campaigning for the acceptance, or at least the toleration, of the Copernican world-system. A biographical sketch which gives prominence to that campaign, with all its philosophical, theological and ecclesiastical consequences, is simply respecting the choice which Galileo made. Nor was it a foolish choice, though it turned out to be an unlucky one. The strategy of the campaign took in not only the promotion of Copernicanism but also fundamental ques-

2   Galileo, from *Serie di ritratti d'uomini illustri toscani*, volume 2
(Florence, 1768).

tions of how science should be approached, questions which needed
tackling if physics was ever to be emancipated from Aristotelian philosophy. His life would be simpler to write if the unhappy consequences of his commitment to Copernicanism could be given a brief
mention and then disregarded, as he taught us to disregard air resistance in the free fall of heavy bodies. But that simpler life would not be

Galileo's. He was not in free fall: he was freely grasping opportunities, responding to difficulties and circumventing constraints. He was not in a vacuum: he was in a sophisticated and learned culture which found many of his views, in the expressive contemporary phrase, 'very impersuasible'. To persuade his contemporaries became his goal, soon after he had caught their attention with his *Sidereal Message*.

The 'Novelist'

There was always the chance, which in 1610 Wotton had to take seriously though we no longer can, that Galileo would turn out to look 'exceeding ridiculous'. *The Sidereal Message* did not declare openly for Copernicanism in so many words, but it gave very broad hints of the way Galileo was thinking. In the dedication to Cosimo II, the Grand Duke of Tuscany, he called the Sun 'the centre of the world' (3: 56). In the body of the booklet he referred three times to a book he hoped to publish on the system of the world (3: 73, 75, 96). The second mention promised full proof of the earthshine for those who thought the Earth must be excluded from the heavenly dance because it is devoid of motion or light: he would confirm by proof and argument that it is wandering and brighter than the Moon. It is only the casualness of this incidental remark that makes it count as a broad hint rather than a declaration. The third mention came in his concluding paragraph, where Galileo said that Jupiter's satellites – to use the name Kepler gave them before the year was out – provide a fine argument for removing an objection to the Copernican system. The objection was that Earth had to carry the Moon round with it as it circled the sun and this seemed to some anomalous, or even impossible. The objection disappeared since there was no doubt that, whether Jupiter circled the Sun or the Earth, it certainly carried four planets with it (3: 95). Nor could anyone continue to claim that all heavenly motions were centred on the Earth, since the motions of the satellites were centred on Jupiter. So, though Galileo did not declare outright that Copernicus had found the true system, what he wrote was very significant: it amounted to transferring Copernicanism from the class of mere calculating devices available to astronomers and putting it squarely among rival views for the true constitution of the universe (see chapter 2). To that extent, it can be said that it was Galileo who made non-astronomers take Copernicanism seriously. *The Sidereal Message* gave fairly unmistakable hints that that was what he intended to do.

Not all readers would take those hints, but none could miss Galileo's definite claim that human vision had gained its first increase in power since the creation of the world; more would now be visible than humankind had ever had access to. It is tempting to say that it could now be *seen* that the Moon was not perfectly spherical, as a heavenly body should be; but Galileo knew very well that it took seeing and reasoning combined to establish that conclusion (3: 62–3; 10: 273). Still, he was quite confident that he could show that the Moon is disconcertingly like our Earth, which almost everyone thought was a unique, fixed body at the centre of the universe, surrounded by the elements of water, air and fire. The defining characteristic of all heavenly bodies, in common estimation, was precisely that they were like nothing on Earth. Yet Galileo was saying that the Moon had mountains and valleys very like those on Earth. The great Hellenistic astronomer, Ptolemy, writing about AD 150, had thought that to treat the Earth as moving was quite laughable.[6] Galileo would soon try to show that to accept that the Earth is moving, is in fact a planet, was the only satisfactory approach. Wotton would not have thought so far ahead, but he could certainly see that the essential split between Earth and heavens was being challenged. Such a challenge, it was reasonable to point out, might well turn out to be ridiculous. A good deal of Galileo's work in both astronomy and physics, both before and after 1610, was devoted to transforming the obviously ridiculous into the ridiculously obvious, a fact which makes it quite difficult for us to understand and appreciate not only the strengths ·and weaknesses of the views he wished to overthrow, but also the successes and shortcomings of his own approach.

There is no doubt, as Wotton perceived, that *The Sidereal Message* announced something very new. If one could use a seventeenth-century English term without misleading people, Galileo could be called 'a novelist': whether by experience or temperament he was fascinated by *new* discoveries, by novelties in theory and practice. His own debt to ancient and medieval thinkers is something which will be touched on as his story unfolds, and he prided himself on continuing the work of Archimedes (287–212 BC), the greatest mathematician of antiquity. But he was certainly an innovator and took equal pride in that. It is not exaggerated to say that his *Sidereal Message* told people that a new age had begun and that the way the universe was studied would never be the same again.[7] In that sense the little book can fairly be called revolutionary. But Galileo's tract, indeed all his work, has to be seen against a wider background if we are not to be beguiled by the easy use of labels like 'the scientific revolution'.

'The Scientific Revolution'

It is still customary to talk of 'the scientific revolution': the label would naturally suggest that at some more or less specifiable period there was such a change in scientific thought and practice that it amounted to a revolution. The challenge would then be to trace the development of later science from that initial revolutionary breakthrough. A seemingly natural refinement of the project would be to concentrate attention on key disciplines, such as astronomy or physics; further refinement might allow that medicine, for instance, should not be forced into a framework designed for astronomy. Further nuances could be introduced: if the period of scientific revolution is supposed to begin, for instance, with Copernicus's *De Revolutionibus* in 1543 and end with Newton's *Principia* in 1687, still one would not have to imagine that Newton, as it were, brought down the final curtain on a completed drama and one could even concede that Copernicus did not make an entirely fresh start. (How much could be credited to 'pre-revolutionaries' without evacuating 'revolution' of all meaning is a topic of continuing debate.) One could also allow for smaller revolutions which would find their place within the one overarching revolution. In other words, 'the scientific revolution' can provide a framework in which to set the antecedents, discoveries, disputes, dead ends and cross-purposes of the sixteenth and seventeenth centuries; in such a framework Galileo could hardly be denied an important role, perhaps even a central one. It could also be said that he claimed such a role for himself: not that he talked of a scientific revolution, but he had equivalent ways of advocating a major reform of physics and astronomy and of scientific method, and he assigned himself a unique place in this reforming movement. So there is a good deal to be said for the usefulness of the label 'scientific revolution', which has been used judiciously by some of the most able historians of science.

But it has its dangers too. There is nothing wrong in tracing the development of ideas, practices, instruments, institutions or worldviews that interest us now. There is nothing objectionable in the persuasion that assignable progress has been made in this or that subject in recent centuries (though philosophers would raise serious questions about the criteria of progress and the status of any claim to knowledge). But a frank pursuit of what interests us now can easily lead to a selectivity which distorts the whole context of what is being examined: what now seems to us alien or pointless is quietly passed over, while other topics or interests are too readily identified with our own. The history of science can, as a consequence, be reduced to a celebration of great scientists, with perhaps minor prizes for best supporting actors. Hindsight can make someone into a precursor or forerunner of someone else, when the simple fact is that, apart from John the Baptist, no one points to someone else who is still to come, though anyone may, as Galileo did, express the serious hope that others will build on his work.

A further danger of the label 'scientific revolution' is that it can tempt us to class people as revolutionaries or reactionaries, winners or losers, progressives or conservatives. Such simplifications are useful, perhaps essential, but they can make us miss the untidy way in which the development of science actually took place. Historians of science are well aware of this danger, but it certainly needs pointing out in any introductory treatment of Galileo's life, because he himself was very impatient with writers more cautious (and less gifted) than himself. He was very much given to grading people on a descending scale which went from free and open-minded observers of nature to servile and bookish dogmatists. Such assessments were sometimes very accurate and often presented with entertaining rhetoric, so anyone studying his life is tempted to concur uncritically with Galileo's admittedly privileged, but nevertheless highly partisan, views on contemporary scientific disputes. One can, of course, innoculate oneself against the disease of adulation by reading some of the considerable literature which sets out to cut Galileo down to size, but the dosage is a matter of dispute and some writers seem to have overdone it. None of this need force us to abandon the useful label 'scientific revolution', though I think it safer to be satisfied with calling Galileo a decisive innovator in disciplines which are important no matter what framework is chosen for the history of science. Great changes were taking place and Galileo was associated with many of them. He was not, of course, seriously involved in them all. He gave up his early study of medicine as soon as he could and his contributions to contemporary medical studies were only indirect or incidental. This false start to his career can serve to remind us that no overall picture of contemporary developments in science can be drawn merely from a study of his life. In this connection it is worth noting that 1543 also saw the publication of Vesalius's great anatomical work *De Fabrica Corporis Humani*. But a decent scepticism about the rhetoric associated with 'the scientific revolution' does not alter the fact that developments in astronomy and physics were of lasting importance. Galileo certainly made physics his own; indeed, one might say that he more than anyone else began to make it into what we now recognize as physics.[8]

But the impact he made on astronomy is rather different. He was undoubtedly a brilliant popularizer of Copernicus's system and of his own telescopic discoveries. He is admired for the way he improved the telescope itself and worked out how to use it to the best advantage. In July 1610 he noticed what appeared to be satellites of Saturn, though his telescopes were not powerful enough to resolve their curious appearance: that had to wait till the work of Huygens nearly fifty years later (10: 474). He discovered the phases of Venus by mid-December and took delight in pointing out that they could not be accounted for by Ptolemaic astronomy (10: 481–3). He tackled the daunting task of working out the periods of Jupiter's satellites with conspicuous success. No one was quicker than Galileo to discern what to make of those puzzling features, the apparent spots on the Sun, which the telescope forced on the attention of European astronomers by 1611. By 1613, with his *Letters on Sunspots* (*Istoria e dimostrazioni intorno alle macchie solari*), Galileo had made a permanent mark on astronomy.

Yet his own interest in theoretical astronomy was curiously limited. He was, of course, a more than competent teacher of astronomical theory, well acquainted with the work of Ptolemy and Copernicus. But when we look for major developments in theoretical astronomy between Copernicus's seminal work and the synthesis (or transformation) of astronomy and physics that Newton gave the world in 1687 we do indeed turn to 1609, the year Galileo made his first telescopes, but it is to Johannes Kepler (1571–1630) that we have to look. In that year Kepler published his *Astronomia Nova*, yet Galileo never seems to have appreciated Kepler's colossal intellectual achievement. It was Kepler who made planetary astronomy truly Sun-centred and geometrically clean, accurate and satisfying in a way that even the marvellous analyses of Ptolemy or Copernicus had never been. It was Kepler who at least tried to assign causes of celestial motions, while Galileo, the physicist, seemed to think that any attempt to provide a celestial mechanics would be premature.[9] Kepler naturally had a proper respect for the observations collected systematically by the Danish astronomer, Tycho Brahe (1546–1601): without the new standards of accuracy, comprehensiveness and error control established by Tycho, Kepler's own work would not have been possible.[10] One might have expected Galileo to have been delighted at Kepler's transformation of Copernicus's system, especially since it was the fruit of daring reasoning tested against precise observation, characteristics of his own best work. Yet he was dismissive of Tycho, who was a matchless instrument-maker and observational astronomer. He usually wrote respectfully of Kepler (even calling him Copernicus's peer) and was glad of his support, yet he ignored Kepler's principal achievements (7: 303, 306). Galileo was put off by the Pythagorean preoccupations which motivated Kepler's researches; understandable though this is, it meant that he never realized that Kepler's theoretical discoveries strengthened his own position very significantly. That is why I shall refer to the heliocentrism espoused by Galileo as 'Copernicanism': he was more or less content with Copernicus's system, despite the serious shortcomings in it which spurred Kepler on to his greatest achievements. It is a classic instance of great minds not thinking alike.

The Turning-Point

Galileo himself knew that the telescope marked a turning-point in his life. Although he was a Tuscan he had been lecturing in Padua since 1592. Padua had been good to him: in old age he looked back on these years as the happiest of his life (18: 209). But the frequent illnesses of the second half of his life began in 1603. In 1604 he had given serious consideration to an offer of employment from the Duke of Mantua, a sign of some dissatisfaction with his working conditions in Padua (10: 106–9). He was chronically pressed for money: although his lecture load was light, he needed to supplement his salary by private teaching and by taking in students as boarders. An indication that he was looking for freedom from routine academic chores came in December 1606, when he wrote spontaneously to the Grand Duchess Cristina of Tuscany to say that his own doctor, the renowned Fabrizio Acquapendente, would happily exchange his chair of surgery and anatomy for employment which would give him time to draw together his research for publication. A republic like Venice could not allow any of its paid servants the leisure to crown an academic career in this privileged way, but an absolute prince such as the Grand Duke could. Galileo made it quite plain that he was not speaking for Acquapendente, whom he had not consulted (10: 164–6). In February 1609 it became clear that when he wrote about Acquapendente he had been thinking about himself as well.

For several years he had cultivated the grand-ducal family, making well-chosen moves in the subtle and well regulated game of client-patron exchanges.[11] A significant breakthrough came with an invitation in 1605 to tutor the young Prince Cosimo during university vacations. The dedication of his manual for the geometrical compass in 1606 was another obvious move (10: 146–7). In 1608 he consolidated his position at court by negotiating the purchase from his friend Sagredo of a magnet of exceptional strength to grace Cosimo's scientific collection;

the time he spent arming the magnet until it could sustain more than twice its own weight was evidence not only of the devotion but also of the unrivalled technical skill that the Medici could command if only they would employ him (10: 205). He even obliged the Grand Duchess in January 1609 with a hopeful horoscope for the ailing Grand Duke Ferdinando I (10: 226–7). His death soon after was the sort of mishap which Galileo, like any astrologer, could explain away as unforeseeable, given the inaccuracy of the data supplied. The succession of his pupil as Grand Duke Cosimo II gave Galileo an opportunity to sound an unidentified intermediary about a uniquely privileged post he had designed for himself. He alluded to three great works he wished to bring to completion and promised various discoveries, with more to come. The reasons for wanting to leave Padua are the ones he had ascribed to Acquapendente. Money was less important than leisure, since he could hardly hope for enough gold to make him stand out among the rich, but he could realistically hope to acquire some splendour from his studies (10: 231–4). It seems that this admirably frank suggestion that a research professorship should be instituted especially for him was not taken up at the time by anyone of sufficient influence. Galileo had resigned himself to renewing his Paduan contract when, in the summer of 1609, everything was changed by his swift reaction to rumours of the new optical instrument or spyglass.

The invention of the telescope has been surrounded by controversy, which began even before Galileo published his *Sidereal Message*. By late September 1608 spyglasses were being made in the Dutch Republic by Hans Lipperhey and Sacharias Janssen, both of Middelburg, and Jacob Metius of Alkmaar: their rival claims (and claims that telescopes had been made earlier) need not be discussed here, especially since the States-General refused a patent on the grounds that the instrument was too easy to copy.[12] The wisdom of that decision was shown by the rapid spread of spyglasses to Paris by the spring of 1609 and to Italy by the summer. Galileo's friend, Paolo Sarpi, had heard a report of the new instrument as early as November 1608. It is not clear when Galileo first heard about it, since his own accounts are hard to reconcile with each other; it may have been as early as May or as late as July 1609. Whenever it was that he heard the first rumour, he had to wait only a few days to have it confirmed by a letter from Paris; the letter was from Jacques Badovere, a former pupil of Galileo's, and was probably written to Sarpi. None of this tells us when Galileo set to work to make his own instrument. A foreigner certainly turned up in Padua with one of the new glasses by 1 August: presumably it was the same person who went to Venice and offered to sell a spyglass to the Republic for 1,000

3   The tower from which Galileo showed off his telescope: a detail from a
view of St Mark's Square, Venice, *Nouveau Théâtre d'Italie . . . sur les desseins
de feu Monsieur Jean Bleau* [i.e. Blaue], tome 1 (The Hague, 1724).

zecchini, on condition that it was not taken apart to discover its secret
(10: 250, 255). Sarpi was appointed to examine its merits, but did not
recommend purchase, doubtless because he already knew that Galileo
could produce something better. Galileo did in fact come to Venice
with an instrument that magnified eight or nine times and showed it off

to several notables from St Mark's campanile. On 24 August he de-scribed it to the Doge as an instrument of great military significance, which would enable the Republic to espy enemy ships two hours earlier than the naked eye could or to see into their fortresses or observe their dispositions from a great distance – a very real advantage in the unlikely event that Venice could keep the method of its manufacture secret. Like any client who knew the rules he offered it free of charge, with just sufficient hints that life tenure at a greatly increased salary would not be unwelcome to him. The demonstration of what was, after all, quite a powerful spyglass, convinced the Venetian senators and Galileo's salary was raised to 1,000 florins a year for life (10: 250–1; 19: 115–17, 587).

This personal coup has brought him some abuse. Even at the time he was accused of passing off as his own an invention which he had simply copied from the instrument entrusted to Sarpi and of swindling the Venetian senators (10: 255, 307). But Venice had an unrivalled intelligence service in European capitals; it is not credible that those who rewarded Galileo knew nothing of spyglasses on their own door-step when everyone else did. It may be doubted whether the senators would be as optimistic as Galileo that the Republic could preserve any notable advantage in their manufacture or use, but from a military point of view the doubling of one professor's salary was a minor expense and they would know that if they did not reward Galileo he could use his new instrument to gain a position elsewhere. The charge of copying cannot be proved or disproved, but it is hard to see why Galileo could not fiddle with lenses at least as effectively as spectacle-makers and hucksters until he hit on a promising combination: to that extent at least he was entitled to claim that he had made use of his knowledge of optics or perspective. It is a curiosity that Kepler had already worked out the theory of telescopes without trying to make one and that Galileo made a great number of effective telescopes without worrying much about theory.

In *The Sidereal Message* Galileo explained that his telescope had a plano-concave eyepiece and plano-convex objective. He did not need to mention that it showed things the right way up, because his readers had no reason to expect otherwise, given the novelty of the instrument. That was just as well: there would be enough insinuations that what the spygglass showed was illusory, without his having to explain away an inverted image. Galileo was not particularly secretive about the manu-facture and use of the instrument. He happily explained to the Jesuit astronomers at the Roman College his puzzling practice of putting an oval ring round the objective lens to stop it down. Odd though it appeared to them, Galileo had discovered that it was easier to grind the central, usable part of a largish lens accurately than it was to grind a smaller one (10: 485, 501). He also had an advantage over most rival instrument-makers in that he knew where to obtain high quality blanks for grinding; the source was Florence, rather than the world-famous centre of glass-making, Venice. Whether or not he had much command of optical theory, he certainly knew how to grind, or have ground under his supervision, lenses which were widely acknowledged to be superior to those made by others. It was not until the 1630s that his telescopes were superseded by Francesco Fontana's Keplerian telescopes with convex eyepieces.[13]

So for two decades no one surpassed Galileo in making the telescope into an instrument effective enough for scientific discoveries of great importance. One can see why, though he always acknowledged it was a Dutch (or 'Flemish') invention, he felt entitled to say in the title-page of *The Sidereal Message* that it had been 'found' by him. One of the most sensible comments (perhaps an echo of Galileo's) was made in 1610 by one of his pupils, John Wedderburn, a Scot. In a book dedicated to Wotton he said that anyone who compared Galileo's telescope with a Dutch one would rightly call him its inventor (3: 158). There is a perfectly acceptable sense in which Galileo was the inventor of *his* telescope, so we can allow his later insistence in *The Assayer* (*Il Saggiatore*) that he was entitled to call it not just his pupil but his child (6: 257–8). The claim is only allowable, of course, because he had greatly improved what had been little more than a curiosity and because he continued to develop the instrument and used it to such astonishing effect in the winter of 1609–10. But he also used it to effect the change he had long been thinking of: to return to Florence with an adequate salary and no teaching obligations.

Return to Florence

The really novel thing in *The Sidereal Message* was the announcement that Jupiter had four satellites. That there are many more stars in the sky than can be seen without the telescope was only an unexpected confirmation of what many had surmised. That the Milky Way is made up of innumerable stars was an opinion that the telescope put beyond doubt, but it was one that had long been held by many, perhaps most. Even the lunar mountains and earthshine had been guessed at by some, including Michael Maestlin and his pupil Kepler. Galileo himself said he had recognized the earthshine many years earlier (3: 72, 117); nevertheless this part of the *Message* was one that many Aristotelians found very hard to accept.[14] But the claim to have discovered four hitherto unseen planets was sensational.

It was on 7 January 1610 that Galileo tacked on to a longish draft letter describing the Moon as seen through the telescope the laconic remark that he had just seen three fixed stars near Jupiter (10: 273–8). It took him three more evenings of observation (and one useless cloudy one) to convince himself that they were not fixed stars but little planets revolving round Jupiter; on the 13th he saw the fourth – Jupiter's other satellites were discovered only much later (3: 83–4). The letter of 7 January may show that Galileo had already resurrected his hopes of resigning his Paduan chair and returning to Florence as a privileged servant of the Grand Duke. (It was also while Galileo was making these historic observations of Jupiter that his mother wrote to Piersanti, Galileo's servant, instructing him to steal a few lenses which Galileo would never miss. She had for some time been exhorting Piersanti to report the household gossip. This is the most striking instance of strained relations between her and her son (10: 268, 270, 279). Family worries of one sort or another provide the backdrop to much of Galileo's work.)

At the end of January Galileo wrote to the Tuscan Secretary of State, Belisario Vinta, to summarize his discoveries and report progress on the printing of his *Message*. He thanked God, who had been pleased to make him alone the first to observe something wonderful which had been hidden from all the ages (10: 280–1). Galileo was never modest about his achievements, but the piety was genuine. That he was also keeping the Grand Duke abreast of his discoveries was natural if he was ever to get what he wanted in Florence. The next step was to dedicate the forthcoming work to the Grand Duke and, better still, to name the new planets after him. 'Cosmic stars' would be a satisfying pun on Cosimo's name, but perhaps 'Medicean stars' would be neater, after Cosimo and his three brothers. (He does not seem to have thought of naming them individually after the brothers. The individual names were actually given by Simon Mayr, who discovered the satellites the day after Galileo.)[15] Vinta replied on 20 February that people might miss the allusion to the family in 'cosmic', whereas 'Medicean' was unmistakable (10: 282–5). Since the booklet was going through the press while the later parts were still being written, the early pages show the correction from 'Cosmic' to 'Medicean'.

On 13 March 1610, the day that saw Wotton dashing off to England the strange news that had come hot from the press, Galileo promised Vinta that he would send his best telescope so that the grand-ducal family could see the new planets for themselves. It was no longer a secret military instrument, as it had been in the autumn of 1609; three days later Sarpi gave a detailed description to a French correspondent (10: 288–90). In any case telescopes, of varying quality, were no longer great rarities. With a full sense of the significance of his discoveries Galileo promised to send the Grand Duke the telescope with which he had made them, asking that it be preserved in its unadorned state. He was already planning to send telescopes to secular and ecclesiastical princes. He claimed to have made more than a hundred (a little later, he says sixty: Galileo was sometimes satisfied with the right order of magnitude), of which ten were good enough to show the satellites. He must have had a very busy winter (10: 279–302). By the end of March the first murmurings of opposition were heard: the satellites were an illusion caused by faults in the glasses and vapours in the atmosphere (10: 309). In April, Georg Fugger, the Imperial Ambassador at Venice, replied to Kepler, the Imperial Astronomer at Prague, the centre of the Holy Roman Empire: many expert mathematicians thought the *Sidereal Message* mere show, a dry and baseless discourse; Galileo was accustomed to decorate himself with others' feathers, like Aesop's crow; he had seen the telescope brought by a Dutchman and copied it, perhaps adding something, which would be easy enough (10: 316). One's respect for Wotton as a prompt, judicious and accurate reporter of the discoveries increases. Not that Fugger was the first to try to turn Kepler against the new discoveries, but he was in any case too late, since Galileo had already sent the *Message* to the Tuscan ambassador in Prague with the express purpose of gaining Kepler's opinion.

Kepler's response was more than generous. By 19 April he had, in less than a week, written Galileo a letter which was a little treatise (10: 319–40): two weeks later he had added to it and prepared it for publication as his *Conversation with Galileo's Sidereal Messenger* (3: 99–126).[16] In it he endorsed Galileo's discoveries, even though he had not had the use of a telescope. No wonder Galileo gave him hearty, if rather tardy, thanks in a letter of 19 August, after keeping him informed in letters to the Tuscan ambassador at Prague (10: 421–3). Incidentally, we know that Galileo referred to his book as *avviso*, which can only mean 'message', whereas the Latin *nuncius* can mean 'messenger' or 'message' (10: 297). The fact that most people took Galileo to mean 'messenger' is largely due to Kepler's adopting that interpretation. Jibes that Galileo wanted to be seen as a messenger from heaven may derive support from his behaviour as a propagandist for Copernicanism, but not from the book's title. In any case, Galileo could put up with a mistranslated title, since he badly needed Kepler's support.

It was no doubt gratifying to be hailed by a friend as a new Columbus (10: 296) and to be approached by cardinals for a gift of a telescope, but it was the endorsement of astronomers that he needed. That endorsement was not easy to secure, as we can see from the case of Giovanni Magini, professor of mathematics at Bologna and a very competent astronomer and geographer. Magini himself told Kepler in May that he thought the discoveries were illusory. This was after Galileo had taken his telescope to Bologna and stayed with Magini towards the end of April: his attempt to show the satellites to his host and assembled guests had been unsuccessful. Kepler's reply in May was to send Magini a copy of his *Conversation*, with a civil little note which did not withdraw his endorsement of Galileo: after all, wrote Kepler, they were both Copernicans. The partisan Magini read the *Conversation*, which did contain incidental remarks of a mildly critical nature, as a rebuke to Galileo. It is true that by July Magini was using a telescope and Galileo was glad to stay with him in September, but it is not hard to see how timely Kepler's earlier support was (10: 342–3, 353, 359, 398, 424). It was that support which helped to change Galileo's life.

On 7 May Galileo wrote a crucial letter to Belisario Vinta. He reported, with characteristic over-optimism, that in three public lectures he had convinced the University of Padua of the genuineness of his discoveries; equally characteristic was his resentment that he had had so little support from other Italian centres of learning. Then he announced triumphantly that he had received the vital endorsement of Kepler. In the rest of the letter Galileo wrote confidently to clinch the proposal he had made more than once about moving to Florence.

His Paduan salary is now 1,000 florins, he tells Vinta. He can more than double this by giving private tuition to foreign gentlemen. All that income is pure gain, if he chooses to finance the running costs of his household by taking in gentlemen scholars as boarders. He has to give only sixty lectures a year. But he wants to be free of boarders and almost free of private teaching, because he needs time for his studies, so that he can complete his books: in other words, he wants to be able to write with no obligation to lecture. The work he proposes to do would be much more useful than lectures, which have to be elementary, could be given by lots of people and do not advance his work at all. So he wishes to earn his salary by his books and by the discoveries he will make for the Grand Duke, who can count on getting value for money. The books he hopes to complete are two *On the System of the Universe*, three on *Local Motion* (a brand new science) and three on *Mechanics*, as well as smaller works such as: *On Sound and the Voice, Sight and Colours,*

The Tides, and *The Composition of the Continuum,* not to mention military engineering and cognate topics. A thousand florins will be acceptable as his salary, but he would like his title to be not just mathematician but also philosopher, since he has studied more years in philosophy than months in mathematics: he will be happy to give proof of his competence in the presence of the most respected philosophers. What Galileo is asking for in this letter will seem a slight matter to Vinta, but it means everything to Galileo (10: 348–53). Vinta replied promptly and favourably; Galileo resigned his Paduan chair on 15 June (19: 125); further correspondence settled details, including a typical request from Galileo for two years' salary in advance. Cosimo made the appointment on 10 July 1610 (10: 400–1).

Galileo's letter to Vinta is remarkable. It shows him taking hold of his life and casting it into a pattern which seemed to him highly desirable, almost essential, if he was to bring to publication the many works, some highly original, which he had largely mapped out. Naturally he was exposing himself to Venetian and Paduan resentment: he had been treated handsomely by the Republic and had used his new fame and increased salary to acquire what he thought was a better post. To accept a generous reward from his patron and then turn it down for a better offer was certainly a slight not easy to forgive. Even at the time, friends thought he was not just discourteous but unwise: Venice was intellectually more lively than Florence and certainly safer for anyone with opinions that might bring him into conflict with the Inquisition. Hindsight naturally sees the move to Florence in September 1610 as overshadowed by the condemnation of the *Dialogue* in 1633: what hindsight cannot see is whether Galileo would have found time to write the book if he had stayed in Padua. In any case, a good deal of the programme which Galileo described to Vinta was actually carried out successfully, and other important works which could not have been foreseen were added to it. To discover not only how he carried out that programme but also how he came to be in a position to draw it up is a principal theme of his biography.

2

Early Life

Galileo was born in Pisa on 15 February 1564. The date is established by two seventeenth-century nativity charts of unknown provenance. The correct date eluded the researches of Vincenzio Viviani (1622–1703), Galileo's first biographer: in the biography be gave 19 February, a date accepted by Galileo's son, though that *was* the date of Galileo's baptism. Viviani's pride in being Galileo's last pupil was expressed in a laudatory commemorative inscription he composed to display on his house. In it he had the Creator consoling Florence and continuing unbroken the city's succession of geniuses by giving the world Galileo two hours before the death of the great Michelangelo: this put the birthday on 18 February. Such hero-worship of Galileo was not initiated by Viviani: it is a significant factor throughout the last three decades of Galileo's life and still has to be reckoned with. Viviani wrote his first draft of the biography in 1654, at which time Niccolò Gherardini also produced a shorter memoir. Neither can be relied on in matters of detail or chronology, but they should not be disregarded completely: at the very least they can direct scholars to profitable areas of research.[1]

I am following the usual English practice of calling Galileo by his first name only. It was also Galileo's own practice when he was annotating opponents' books with a view to publishing a reply: he often jotted down remarks such as 'Galileo never said that.' He is the last of a few great Tuscans such as Dante, Leonardo and Michelangelo to have been extended this compliment by widespread consent.

4   Vincenzio Viviani, Galileo's first biographer, from *Serie di ritratti d'uomini illustri toscani*, volume 2 (Florence, 1768).

The Galilei family could trace its ancestry back to Giovanni Bonaiuti in the thirteenth century. Giovanni's great-grandson, Galileo Bonaiuti, adopted the fashion of repeating his first name as a surname, so that Galilei replaced Bonaiuti as the family name. This first Galileo Galilei, the elder brother of the scientist's great-grandfather, died about 1450 and his effigy can be seen on his tomb in Santa Croce in Florence. He had been a very successful doctor and an influential professor and administrator in the University of Florence. He had also held high public office in the Republic, as did several of his descendants (19: 15–16). This ancestry is important. By the middle of the sixteenth century the family was in reduced circumstances but it was still conscious of its pedigree. Galileo would not have to wait for fame to give him access to influential people: he could follow the normal route of attaching himself to increasingly powerful patrons. Though deferential to rank, and even sycophantic where custom so required, he was eventually to move with reasonable ease in the higher reaches of Church and State, with only occasional spasms of insecurity. On the title-pages of his published works he styles himself *nobil Fiorentino*, a noble Florentine: being born in Florence's satellite of Pisa did not deprive him of this proud title.

His father, Vincenzio (1520–91), married the Pisan Giulia Ammannati (1538–1620) on 5 July 1562. Part of her dowry was made up of linen and woollen cloth since Vincenzio, it seems, was settling temporarily in Pisa to support his family as a cloth-merchant, a traditional and respected occupation which had done much both to build the prosperity of Florence in its great days and to finance many of its artistic treasures (19: 17–20, 26–31). Much more significant for Galileo, however, was his father's chosen profession as a musician. Vincenzio has his own respectable little niche in the history of music. He had studied under Gioseffo Zarlino at Venice and had conferred frequently with Girolamo Mei in Rome. As a music-master he would give lessons at home, just as Galileo was later to supplement his professorial income by tutoring in applied mathematics. Vincenzio lived to see Galileo an accomplished lutenist and a fine singer. He cannot have guessed that the mathematical side of musical theory was to influence his son's achievements in new fields, but he would not have disguised from his son his dissatisfaction with the theories he had been taught. Galileo is usually seen above all as a subverter of established views. Whether his later eagerness to take on the world of learning is something he got from his father need not be determined. What he did get from him appears incidentally in his mature scientific work.

In 1572 Vincenzio Galilei returned to Florence, leaving his wife and Galileo at the house of her relative, Muzio Tedaldi, who was a business contact. Tedaldi's accounts show Galileo paying 5 lire monthly to 'the master' (19: 26–31), presumably a Latin teacher, tentatively identified by Favaro as Jacopo Borghini.[2] In January 1575, after Galileo had been restored to his father in Florence. Tedaldi hoped Galileo was growing in virtues and letters, which tells us little except that Tedaldi was very affectionate towards the family, as indeed he says in so many words

5   Niccolò Gherardini, the author of an early biographical note on
Galileo, from *Serie di ritratti d'uomini illustri toscani*, volume 2 (Florence,
1768).

(10: 18–19). Virginia Galilei was born in 1573, preceded by Benedetto
and followed by Anna, both of whom are merely names to us.
Michelangiolo was born in 1575 and the family was completed by the
arrival of Livia in 1578 (19: 15–16). A couple of references to 'our Lena'
are sometimes used to provide Galileo with another, otherwise unknown, sister, Elena (10: 60). Little is known of Galileo's mother beyond
the fact that in her widowhood she was a trial to her children, especially
Galileo.

Gherardini says that Galileo studied Latin near home in the Via dei Bardi (19: 635); Viviani makes this study last some years, after which he studied logic under a Vallombrosan priest (19: 602). Though the chronology is not clear, the Vallombrosan episode is documented in a seventeenth-century manuscript written by a Camaldolese friar, Diego Franchi, but now available only in a book printed in 1864. Franchi says that Galileo was a Vallombrosan novice and made the first exercises of his admirable genius in the Camaldolese school of Vallombrosa. But his father, under the pretext of bringing Galileo to Florence to cure him of serious eye-trouble, diverted him from the order. The Vallombrosan connection is also mentioned in contemporary notes on a court case of 1590–2 in which Galileo was a witness. Repeated references are made to his having been a monk or friar; according to the hostile annotator he was expelled or unfrocked, though perhaps this means no more than the 'spoiled priest' label used even in living memory of boys who decided not to continue to ordination. A mention of Galileo's having been a friar at Santa Trinità (in Florence) suggests that he continued to study with the order there, after his stay at Vallombrosa (19: 46). From all this it seems likely that Galileo received part of his pre-university training at Vallombrosa somewhere between 1574 and July 1578, when he was back with his father, and that he continued these studies for some time with a Camaldolese monk in Florence (10: 21). Galileo also learned some Greek, as an exercise in translation shows; he seems to have mastered it sufficiently to use it occasionally later in life.[3] The early biographers, perhaps over-apologetically, agree that Vincenzio could not afford a better early education for his son.

From Galileo's very few references to his childhood and youth a couple may be picked out. In 1577 there had been a spectacular comet and Galileo recalled in a later controversy that he had seen it continually: unfortunately he does not say where he was at the time (6: 314). While making notes for the same controversy he compared his adversary to an awkward customer (6: 160). It is a harmless fancy to see in this passage the young Galileo looking after the shop and muttering at the nuisance who takes material out of the shop to try to find the tiniest fault so that the price can be beaten down, regardless of all the diligence, patience, time and trouble that has gone into making the rest of the bolt of top quality. The vignette is completed by the not entirely relevant remark that the stuff will be muddied and ruined before evening anyway.[4] If Galileo ever did help his father in business, he would not have subscribed to vapid notions of the customer's always being right.

Student at Pisa

As early as 1578 Vincenzio was searching for ways to support his son at the University of Pisa. At first he hoped to obtain a bursary at the Sapienza, a small college for needy Tuscan students; failing that, Tedaldi was willing to give Galileo lodging (10: 19–21). It was not till 5 September 1580 (or perhaps 1581) that Tedaldi's offer was taken up and Galileo was matriculated in the arts faculty as a student of medicine (19: 32).[5]

The University had been the pride of the Medici Dukes of Florence (Grand Dukes of Tuscany from 1569) since its refoundation under new statutes in 1543. (The University of Florence may be disregarded at this period since it was more or less in mothballs.) Not even Medici support could win the University the international renown still accorded to Bologna or Padua, but that did not make it a backwater. It could not hope to attract and hold many lecturers of outstanding ability but it was a respectable institution of medium size, with upwards of 600 students: about half of them came from outside Tuscany and there would usually be a few dozen from abroad.

It was divided into three faculties: theology, law and arts. Theology was the smallest and least important, though university-based theology was to expand considerably during Galileo's lifetime as part of the Catholic reform or Counter-Reformation which followed the Council of Trent (1545–63). That the Church already had extracurricular ways of influencing all the university's faculties hardly needs saying as a preface to Galileo's academic career. The most important faculty was that of law: in this both civil and canon (Church) law were taught, so it was a gateway to careers in State and Church. Law accounted for roughly three-quarters of the students, with most of the remaining quarter in the arts faculty. One must not be misled by English or French models, where arts provided the undergraduate foundation for the higher faculties of theology, law and medicine. In Pisa the arts faculty was really the medical faculty, served most importantly by philosophy and a few subsidiary subjects, mathematics being one of the least of these; the arts faculty had as many teaching posts as law. So when Galileo matriculated to study medicine he was embarking on one of the two obvious ways open to young Tuscans of preparing for lucrative professional but not ecclesiastical employment.

Galileo himself (with the help of his early biographers) has done as much as anyone to persuade historians that the University was an obstacle to scientific progress at this period, though it must be allowed that there is solid evidence to support his criticisms. Yet some caution is needed: it must not be assumed that what we shall see of Pisan Aristotelian physics and cosmology was typical of medical studies there. The University had had a botanical garden from the time of its refoundation and had been served by some distinguished physicians. This suggests that Aristotelianism's contribution to medical studies was not the sorry story that will emerge from Galileo's controversies with Aristotelian natural philosophers.

What we know of his student days is very meagre, but we do know the outcome: he was one of many students who left the University without a degree. That was in 1585. How much attention he gave to the prescribed lectures in medicine, philosophy and mathematics cannot be discovered. He may have been introduced to Plato, since Francesco de' Vieri (Verino) still held the new chair in Platonic philosophy that had been estabished in 1576. It is a safe assumption that he must have followed the lectures of the philosophers, Francesco Buonamici (or Buonamico, died 1603) and Girolamo Borro (or Borri, 1512–92), yet his later familiarity with their ideas could have come from their published works. Since the Aristotelian philosophy taught by Buonamici, Borro and countless other lecturers all over Europe was to be the target of Galileo's severest criticisms, a preliminary sketch of some of its main features will be helpful.

Aristotelianism

Aristotelian philosophy covered a great range of subjects. Some, such as ethics, are still very much part of philosophy syllabuses. Indeed, Aristotle's own moral philosophy is still highly regarded. In logic he was long considered to have exhausted the subject (allowing for some important contributions by later authors); even now, after the great expansion of the subject, his treatment of the part he dealt with is recognized as the pioneering work of a master. Other subjects treated by Aristotle have long since emancipated themselves from philosophy and are now classed as scientific specializations. In some of these subjects, such as biology, Aristotle's ideas are treated as creditable contributions to a discipline which bears at least a strong family resemblance to modern biology. With Aristotle's physics, however, one has to wonder whether the subject shares much more than the name with what we now expect to find in textbooks of physics. At the same time, Aristotle did deal in his own way with topics such as falling bodies or projectiles, so it may be too sweeping to say that physics started only after

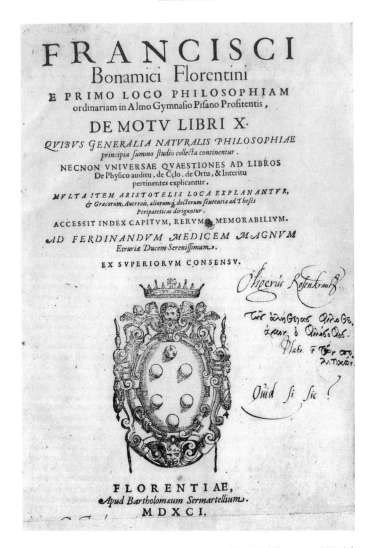

6   The title-page of the massive *De motu libri X* (Florence, 1591) by
Francesco Buonamici, one of Galileo's Aristotelian teachers and colleagues
at Pisa.

Aristotle's ideas were abandoned. Whether Galileo (and others) were
the founders of physics, or whether it is more accurate to say that the
seventeenth century saw the birth of a new physics, are questions
underlying much of the discussion of Galileo's own work. In this short

biography it will be sufficient to advert occasionally to a single question: how much of a 'novelist' was Galileo?

The Aristotelian philosophy taught at universities was not a monolith, even in Galileo's student days, when many Aristotelians were unimaginative continuators of a dominant but tired tradition. Nor was it entirely immune to influences which are usually seen as alien to it. Buonamici, for instance, was a typically ponderous Aristotelian: his work on motion exceeds a thousand folio pages and weighs three kilograms. Yet he seems to have considered himself something of a humanist, since he quotes classical authority for words which purists might proscribe as scholastic barbarisms and scatters Greek phrases throughout his Latin text.[6] Then there was the engaging scholastic tradition of expounding a variety of hostile opinions prior to refuting them. In this way students would quickly acquire a nodding acquaintance not only with some of the opinions of Plato, for instance, but also with more subversive currents of thought, such as ancient scepticism or atomism. But the Aristotelianism of the late sixteenth century centred on the views I shall mention here. Galileo could not help becoming familiar with such views if he went to any lectures at all at Pisa, or even if he read any typical works of philosophy.

Two things should be borne in mind in this context. The first is that there is much more to Aristotle himself than his physics. Anyone who knew no more of his philosophy than what can be gleaned from criticism of his physics by Galileo and his contemporaries would wonder how Aristotle ever came to dominate the intellectual life of Europe. The simplest answer is that, from the thirteenth century onwards, he seemed to provide a comprehensive body of knowledge and an accompanying account of human processes of understanding, which needed only a few crucial corrections to accommodate everything known and knowable by human reason while still allowing a proper crowning place for what came from the Christian faith. Whether a genuine Aristotelianism was reconcilable with the Christian faith was admittedly a matter for continual dispute. (Borro, an avowed slave to Aristotle, had trouble with the Inquisition.) Whether Aristotle would have recognized his own views in what commonly passed as Aristotelianism in Europe's universities was also open to question. Modern scholars of Aristotle rightly point out that his comments on motion, for instance, were not the tidy 'laws' which hindsight finds easy to extract and criticize. But such 'laws' are a legitimate simplification of how Aristotle was generally understood by his followers. Galileo was only one of many who attacked some aspect or other of Aristotelianism. His adversary was to be not so much Aristotle himself as the domesticated Aristotle of university philosophers (like Borro or Buonamici) who showed no fundamental misgivings over the weakest part of the 2,000-year-old Aristotelian body of knowledge, its physics. Galileo would always be ready to allow, indeed insist, that if Aristotle had had the benefit of later knowledge he would have changed his opinions. The second thing to bear in mind is that Galileo was by no means the first to criticize Aristotelian physics. The criticism had started centuries before and reached its high point in fourteenth-century Oxford and Paris. The achievements of medieval philosophers, particularly their sophisticated treatment of various kinds of change, including motion, have been properly appreciated only since the monumental pioneering work of Pierre Duhem opened up this topic to serious scholarship. It is not too much to say that the work of twentieth-century scholars has transformed our understanding of medieval science.[7] But whatever the achievements of medieval thinkers, some of which find parallels in Galileo's writings, it remains true that Aristotelianism had yet to be unseated from university chairs of natural philosophy at the time Galileo became a student.

The fact that it was philosophers who taught the subject then known as physics, along with the fact that to a modern reader what they discussed does not look at all like physics, shows that there really was a major shift in thought in Galileo's lifetime: that shift is a principal theme of his intellectual biography. Aristotle's treatment of how bodies move was one of the characteristic determining features of his world-picture or cosmology. Heavy bodies move downwards in a straight line towards the centre of the universe, which is the centre of the Earth; light ones, that is, bodies which have positive lightness, move away from the centre, again in a straight line. This is the key to the doctrine of the elements. Earth goes straight down; fire goes straight up; air goes up because it is light, but not as light as fire; and water goes down, because it is heavy, but not as heavy as earth. So we have earth, surrounded by water, then air and finally fire: these four spheres make up the elementary world below the Moon. The only bodies that are neither heavy nor light are by the same token non-elementary: they are heavenly bodies with their own appropriate motion, which is circular round the centre of the universe. Circular motion is appropriate to heavenly bodies since no change has ever been observed in the heavens. All we see is endless repetitions of the same patterns of movement, but there is no trace of the generation and decay which is the mark of our elementary world. The heavens must be made of a fifth element (quintessence), an imperishable, incorruptible substance. Since the quintessential heavens are completely different from the Earth and its surrounding elements, there could be no thought of treating all motions in the universe as subject to the same laws. As far as local motion on or near the Earth was concerned, Aristotle was content with principles that were more or less satisfying to commonsense, at least until subjected to a serious examination.

There was no need to explain why a body was at rest in its natural place: that was where it was supposed to be, so it could not be expected to move from there unless forced to. Physics was the study of nature: central to it was the study of natural motions, the study of how bodies return to their proper places. It was motion, not rest, that needed an explanation. There were, of course, also motions that were not natural: these were forced or violent motions. Things like chairs or carts or spears did not move of their own accord: to move them took effort, which was to be expected because they were being moved from their natural rest; violent motions merited incidental attention. But even with natural motions like the free fall of heavy objects, the resistance of the medium had to be considered, since a stone, for instance, obviously falls more quickly through air than through water. In fact, Aristotle took heavy bodies to fall with speeds proportional to their weights in a given medium. He also took the speed of fall to be inversely proportional to the resistance of the medium, though what Aristotle had in mind is more faithfully captured by Galileo's terminology: the more subtle the medium, the faster the body falls; the crasser, the slower.

That such ideas were highly vulnerable to criticism had long been obvious and we shall see that Aristotle and his followers were at their least convincing on the forced motion of projectiles. Yet even the brilliant discussions of philosophers at Oxford and Paris in the fourteenth century had not led to a dismantling of the Aristotelian world-picture. The comprehensiveness and satisfying meaningfulness of the Christian world-view, which incorporated Aristotle's legacy, left plenty of room for discussion, discussion which could more or less accommodate competing accounts of falling bodies or projectiles with no urgent awareness that a thorough survey would show important parts of the structure to be ramshackle. In any case, most people would not choose to vacate even this rickety building until they could see hope of a better habitation. That Galileo himself was seriously dissatisfied with what he learned of Aristotelian philosophy at Pisa seems clear enough. But there is no reliable evidence that he articulated any criticism of Aristotle's views on motion while he was a student, unless we attach to this period his much later claim that seeing hailstones of varying sizes hitting the ground together made him doubt whether they had fallen from different heights (7: 331).

Copernicanism

A motionless Earth at the centre of the universe was an essential feature of Aristotle's world-picture. Yet nearly four decades had passed since Copernicus had revived the idea of Aristarchus of Samos (*c*.320–250 BC) that the Earth is a planet which rotates daily on its axis and revolves annually round the Sun. Copernicus had in fact done more than revive an idea: he had published a fully worked out system which he put forward as preferable to the so-called system of the Hellenistic astronomer Ptolemy. That Copernicus's arrangement of the planets, including the Earth, is truly a system is perhaps the most striking thing about it: everything in it hangs together. In contrast, the Ptolemaic approach in the *Almagest* and in the practice of many of those who followed Ptolemy was more a collection of mathematical techniques to serve positional astronomy: it could not justify a commitment to a unique arrangement of the planets, still less to any ratio of their orbits which could yield an estimate of their distances from the Earth. It was common for people to accept that somehow or other the refinements which Ptolemaic astronomers found necessary for accurate prediction could be fitted in to a general picture derived from Aristotelian philosophy, where each planet had its own crystalline sphere. There was not usually felt to be any pressing problem about combining a simplified picture of nested heavenly spheres with the complicated geometrical devices found necessary by astronomers in their professional work. It had long been accepted that the astronomer's task was to make accurate predictions of the positions of heavenly bodies and that to do this he had to make use of quite complicated combinations of circular motions to account for celestial motions, which in themselves were thought to be uniform. The devices used by astronomers, such as eccentric circles or circles (epicycles) carried round by other circles, were commonly called hypotheses, but the use of this word did not imply that there was anything in the heavens corresponding to a particular set of hypotheses. It was sufficient if the hypotheses used reproduced the planetary positions with sufficient accuracy: there was no need to make the claim that no other set of hypotheses could do the job equally well. This approach to astronomy can conveniently be called instrumentalism: according to it the hypotheses used by the astronomer are calculating devices which need not be true provided that they do the job they are designed for, namely facilitate accurate predictions. This assessment of the status of astronomical theories was widely accepted at the time Galileo enrolled at the University of Pisa, though some Ptolemaic astronomers, such as the Jesuit Clavius, were dissatisfied with it. Most astronomers, of course, would be instrumentalists only in their professional work: they, like almost everybody at the time, were realists about the Earth itself: it really is motionless; to think otherwise would be ridiculous.[8]

It was the philosophers who had the task of describing what the heavens are really like: it made no difference to the work of a positional astronomer whether or not heavenly bodies are made of a special celestial substance, as philosophers insisted. It would make a great deal of difference if an astronomer wished to claim that heavenly motions in reality are not uniform and circular, because then he would be challenging the philosophers' sole right to decide questions of reality. But no one had yet thought to make that challenge. Copernicus, however, did make a different challenge. He put his system forward as the true system of the universe. He was quite serious about treating the Earth as a planet and making the Sun the centre of all planetary motions (though he fell short of his aims in that the centre of the Earth's orbit, which did not coincide with the Sun, was the actual centre of his system). One would expect such a challenge to cause a tremendous stir among Aristotelian philosophers. The main reason that no such thing happened is a curiosity of history: Copernicus was dying in 1543 and his book was seen through the press by Andreas Osiander, who added a preface which warded off any attack on the book by the simple expedient of saying that Copernicus was putting forward his ideas only as hypotheses (meaning mere calculating devices). The preface was very effective. Copernicus could be praised as another Ptolemy by astronomers, who used his hypotheses happily when it was convenient. Astronomers could, of course, spot the contradiction between Osiander's preface and the body of Copernicus's book, so they had to decide for themselves. Very few, in fact, followed Copernicus's realist approach. Copernicus, after all, had not by any means solved all the problems which are involved in treating the Earth as a planet. Most non-astronomers would be satisfied by Osiander's reassurance, if they even needed it. Not many people would need reassuring that the Earth is motionless and they were habituated to the idea that astronomers had a licence to play with all sorts of odd devices. In any case, as can be seen from Buonamici's massive tome, no more than a routine application of familiar principles seemed needed to show that there were insuperable objections to any idea that the Earth really is moving.[9] So it is not surprising that by the time Galileo enrolled at Pisa Copernicus's book had made no great stir in university circles. It was to be Galileo himself who eventually made the stir, but Copernicanism was not widely discussed as a live issue when he was a student, or even when he was a professor: one could say that there was as yet no such thing as Copernicanism.

Mathematics

One of the minor subjects taught at Pisa was mathematics. The chair was held by Filippo Fantoni, a Camaldolese monk of no great academic distinction. He was in the Pisan tradition, which seems to have leaned to astrology rather than the serious mathematics pursued elsewhere in Italy. Even so, it would have been possible for Galileo to get his first taste of mathematics from the rather elementary courses taught by Fantoni. The 1582 course (repeated in 1584) on the fifth book of Euclid covered a topic, proportions, which was a vital tool in all Galileo's mature work in physics. Yet there is no record of Galileo's having attended any of Fantoni's lectures, so there is nothing to throw doubt on the assertion of his early biographers that his mathematical initiation did not take place in the University.[10]

The inconsistency in details in the versions given by Viviani and Gherardini is perhaps no more than we should expect in reports of an old man's recollections of how his career started, though we also have to allow for the biographical tradition of giving a great man in his early years the help of an established master, as Cimabue is supposed to have plucked Giotto from shepherding to be a painter.[11] The core of their stories is that Galileo managed to worm his way into hearing or overhearing the lectures on Euclid which Ostilio Ricci gave to the pages at the Tuscan court. Ricci (1540–1603) is said to have been a student of Tartaglia's, so he linked Galileo to one of the revivers of the study of Archimedes. Galileo was immediately fired by an interest in mathematics. His father, as a distinguished theoretician of music, was no enemy of mathematics but wanted his son to qualify in medicine, a sure means of support compared with mathematics, or with music for that matter, as he knew only too well. Gradually, with a certain amount of connivance between Vincenzio and Ricci and with Galileo indulging in occasional role-playing as a diligent medical student, it was agreed that he was serious about mathematics and should be allowed to pursue the subject. The anecdotes are pleasing and can be taken as showing that it was Ricci who introduced Galileo to Euclid, perhaps as early as 1583, and to Archimedes, perhaps by 1585. That would be more than sufficient to make Ricci a very significant figure in Galileo's education, but he did more: he introduced his pupil to mathematical techniques of

7   Lodovico Cardi da Cigoli, from *Museo fiorentino: Serie dei ritratti degli eccellenti pittori*, volume 2 (Florence, 1754).

measurement and to the study of perspective. One of Galileo's fellow-pupils was Lodovico Cardi da Cigoli (1559–1613), who became a much sought-after painter. At the time of the telescopic discoveries he was one of Galileo's correspondents in Rome. As a good and loyal friend who happened to be painting the Virgin Mary with the Moon at her feet in Santa Maria Maggiore, he introduced the lunar features discovered by Galileo (11: 449). A mastery of perspective was a signal help to Galileo in making those discoveries, as Cigoli remarked. Galileo himself could have been a painter, as is evident from his drawings of the Moon. Later in life he was often deferred to as an outstanding judge of painting. As it was, Ricci set him on the way to a career in applied mathematics.[12]

There is a clue to how the mature Galileo saw his younger self at this stage in a letter he wrote decades later about Bonaventura Cavalieri (14: 36). Cavalieri had been introduced to mathematics by Benedetto Castelli (Galileo's own pupil): that took only a few days. After that Cavalieri had little need of help from others (except occasionally from Galileo), since those who need teachers in mathematics never excel, for natural ability is worth more than a thousand tutors. Substituting Ricci for Castelli and Galileo for Cavalieri, we have Galileo's own mature perception of himself as he was on the threshold of his career. We shall see in a moment that he even found someone to give him the occasional help that he supplied to Cavalieri. Like Cavalieri, the young Galileo was made for mathematics. The study of medicine had never had much appeal for him: he left the University in 1585 without a degree and returned home to continue his mathematical and other studies.

It seems certain that during the next few years he helped his father with his acoustical experiments: some of these found their way into Galileo's last and greatest book. Thus it was at home that he was introduced to experimental science. Carugo and Crombie point out that the constellation of music, perspective painting and mechanics fitted happily into the search for a rational science of nature in Aristotle's tradition. The mathematical arts, which Galileo was only beginning to master, were to change that science of nature decisively.[13]

One of the handful of stories from the history of science in popular circulation is that of Archimedes leaping from the bath and running naked down the street, shouting 'I've found it.' What he had found was the solution to the problem of whether a crown made for King Hiero was pure gold. Galileo gave his own independent studies an appropriate start by taking up this problem in a very brief Italian work written in 1586 (1: 214–20). Dissatisfied with the crude procedure traditionally attributed to Archimedes, he made use of his own study to reconstruct what must have been the true reasoning of the great mathematician. He described a hydrostatic balance and explained how to weigh pure metals in air and water and thus proceed to find their proportions in alloys. One characteristic feature of this clear little essay was his method of making precise measurements along the arm of the balance by winding very fine brass wire round it and counting the number of

8   Clavius, from Isaac Bullart, *Académie des Sciences et des Arts*, tome 2,
livre 2 (Amsterdam, 1682).

turns between the points to be measured. Galileo could hardly have
introduced himself more neatly: here was a disciple of the great
Archimedes, tackling problems in the real world and using his hands to
measure nature as exactly as possible.[14] He was even closer to the

— 37 —

Archimedean ideal in work he did in 1587 and 1588 on the centre of gravity of solids (1: 187–208). It was this work which was to gain him the attention of mathematicians and open up to him an academic career.

Clavius and the Jesuits

The obvious career for Galileo was that of a university lecturer in mathematics. He kept an eye on posts which fell vacant and was not afraid to aim high. Sometime in 1587 (or early 1588) a friend recommended him to a vacant chair at Bologna as an esteemed pupil of Ricci, very well-versed not only in mathematics but also in humanities and philosophy, with some teaching experience, namely a public lectureship in Siena and lectures to many gentlemen in Florence and Siena.[15] This reference would not have seemed very impressive to the Bolognese senators: the public lectureship at Siena might sound promising, but nothing is known of it and it may have been much less than it sounds. The lectures to gentlemen in Florence may have been more or less private, though Galileo is known to have given two Italian lectures to the Florentine Academy on the form, location and size of Dante's hell (9: 29–57). His easy use of diagrams to illustrate quite detailed geometrical arguments and his casually displayed familiarity with what Dürer and others had written about the proportions of the human body must have rendered his vindication of the interpretation of Antonio Manetti, an Academician, quite impressive. But it is not known when the lectures on Dante were delivered; 1588 is a probable date. In any case, his achievements were as yet hardly enough to win an important chair without influential backers. Since he visited Rome, probably towards the end of 1587, we may assume that his purpose was to solicit recommendations.

It was natural for him to visit the leading Jesuit mathematician, the German Christoph Clau (1538–1612), known as Clavius, who had taught in the Roman College since 1565. Two of Galileo's later controversies were with Jesuit mathematicians; it was the Jesuit Cardinal Roberto Bellarmino (Anglicized as Robert Bellarmine ever since his controversy with James I) who in 1616 told Galileo officially that Copernicanism could not be held as true; and Galileo's friends blamed Jesuits for his condemnation by the Inquisition in 1633. But those are not the only reasons which make it important to understand what Jesuits thought and to notice what they had in common with Galileo, as well as what led them to differ.

The Society of Jesus had been founded by Ignatius of Loyola (1491–

1556) only in 1540, but it was already a very important and far-flung educational institution (though education was by no means the only work of this remarkable order). Its schools and colleges in many countries taught a full syllabus of humanities, philosophy and theology. After years of consultation, in which Clavius had played a part, the Society produced in 1586 a book printed for its own members, entitled *Ratio Studiorum*, detailing the pattern of studies for all Jesuit schools and colleges. This was to be revised in the light of the experience of Jesuits in different countries. The definitive version was printed in 1599; that text provides essential background for understanding the Jesuits with whom Galileo came in contact over philosophical, theological and scientific questions. Two things are directly relevant to Galileo's own interests. One is the problem of which opinions could be taught safely, that is, without danger of undermining the Catholic faith. The other is the place of mathematics and philosophy in Jesuit education, where Clavius is the key figure.

That the Society should be very concerned that all its teachers should be orthodox is to be expected. What is perhaps surprising is that there was considerable disagreement among Jesuits both in theology and philosophy about what was safe and what was not, a disagreement evident in many documents from 1560 onwards. So, for instance, in theology, we see Francisco Suarez having to defend himself by saying that, though his method was to some extent new, the opinions he followed were traditional; in philosophy we find Benito Pereira, whom Galileo thought worth reading, accused of being an Averroist, that is, over-influenced by the great Islamic commentator on Aristotle. Pereira, in fact, had no time for those who idolized Averroes, but he would not praise those who thought all should flee from him as a mental plague. This open-mindedness was coming to seem unacceptably fuzzy.[16]

One of the principal things the *Ratio* was meant to settle was the difficult question of *the choice of opinions*. This has some similarities with a principal question of the philosophy of science: how does one choose between rival scientific opinions? The similarities, however, do not alter the fact that for every Jesuit there was a higher court of appeal than philosophy or science, namely the teaching of the Church. No opinion which conflicted with the teaching of a Church council, for instance, or with the clear meaning of scripture, could be countenanced. (This was a boundary which Galileo too always accepted, if properly understood.) But that still left room for all sorts of disagreement among Jesuits. We have seen that the truth or falsity of the Copernican system of the universe was not one of the questions much agitated at the time the *Ratio* was being composed. Clavius showed no signs of alarm when he gave reasons for rejecting Copernicanism, though he thought it physically absurd and in conflict with scripture. But it is still a good example of just the kind of tricky position which might find supporters and so become controversial. If it did it might not be decided easily, unless of course it was settled by authority.

Clearly it would be of enormous help to an international teaching organization whose prime concern was orthodoxy if there could be a reasonable amount of uniformity in its educational institutions. The choice the Jesuits made was to follow St Thomas Aquinas as a general guide in theology and Aristotle in philosophy. Anything in Aristotle contrary to the faith was, of course, to be rejected and suspect interpreters were to be treated with great caution. There should, for instance, be no separate treatise in the philosophy syllabus on the opinions of Averroes; if something good was used from him there should be no praise, but, if possible, lecturers should show he got it from somewhere else. Lecturers should not introduce new opinions without consulting their superiors; those who were prone to novelties or of too free a mind should be removed from teaching.[17]

It would be very easy to use these and similar snippets from the *Ratio* to make Jesuit teaching of philosophy seem little more than a wooden repetition of a theologically domesticated Aristotle and there is no doubt that the *Ratio* remained in force without revision long after its provisions were obsolete. But at the time these provisions were introduced they seemed sensibly cautious; a fair assessment would compare them with contemporary university syllabuses rather than with the writings of pioneers like Galileo, who turned out to be successful. They would also come out reasonably well if compared with contemporary occultist or magical writings. On the question of introducing novelties in philosophy, it does not seem that the *Ratio* was bound to make things cripplingly difficult for Jesuits who took an interest in new discoveries. It is quite possible that, if there had been no official condemnation of Copernicanism, the *Ratio* itself would not have been any great obstacle to Jesuit acceptance of the new system, as can be gathered from the sympathetic attitudes of Christoph Grienberger and Paul Guldin, pupils of Clavius. We shall see, however, that the implementation of the *Ratio* was to become increasingly cautious just at the time that the first telescopic discoveries were being discussed. That early in his career Galileo himself was happy to learn from Jesuit authors, including Aristotelian philosophers, is no longer in dispute. The only dispute is about *when* he applied himself to their works and whether he had access to manuscript sources in addition to printed ones; that is a question which will be touched on later. What Galileo's own life makes clear to us is that the Jesuits had espoused Aristotelianism just when it was about to suffer sustained attacks from many quarters. One of the most effective attacks was to come from Galileo.

One strength of the Jesuit syllabus in philosophy, which would have appealed to the young Galileo, was that it made some provision for mathematics. That it did so was the achievement of Clavius, who persuaded the Society that a basic grasp of mathematics was very useful for a proper understanding of philosophy. The way it was incorporated into the syllabus went some way to breaking down the usual compartmentalization whereby philosophers felt free to marginalize or even ignore mathematics. Clavius was looking to improve the teaching of Aristotelian philosophy by encouraging sensible modifications: he had no wish to subvert Aristotelianism, still less to accept Copernicanism as the true system of the universe.[18]

In astronomy he was an intelligent follower of Ptolemy, cautious but still ready to accept changes that were sufficiently supported by the evidence. He was not satisfied with the instrumentalist view that the phenomena can be accounted for by fictitious circles that in no way correspond to the actual causes of what we see, nor with the claim that, no matter how successful a given set of hypotheses was in accounting for what is observed, there could always be a better set, as yet unknown to us. (This instrumentalism may sound to modern ears like an acknowledgement that even the best scientific theories are provisional and likely to be superseded. In fact, as we shall find later, the similarity is misleading.) Clavius was a professional astronomer who wanted to have some means of choosing rationally between rival hypotheses as explanations of reality. He recognized that instrumentalism, for all its attractions, undermined realism not only in astronomy but also in all our attempts to discover real causes from observed effects. His challenge to sceptical instrumentalists was: 'Show me better hypotheses, or accept what I put forward.' What he required of astronomical hypotheses was they should fit the observed facts, generate accurate predictions and not conflict with natural philosophy. This sane and sober philosophy of science was not sufficient to dislodge instrumentalists from their position: indeed, instrumentalism is still an influential philosophy of science. Later we shall see how Galileo developed arguments used by the Ptolemaic realist Clavius to support the truth of Copernicanism and how the authorities of the Catholic Church took refuge in the instrumentalism which Clavius attacked.[19] Clavius's philosophy of science is open to serious criticism; so (perhaps as a consequence) is Galileo's. But so is every philosophy of science.

The Roman College itself was seen as the Society's educational model of how to teach theology and philosophy. Although mathemat-

GVIDIVBALDI
E MARCHIONIBVS
M O N T I S
MECHANICORVM
L I B E R.

TOLLERET
QVIS
SI
CONSISTERET

P I S A V R I
Apud Hieronymum Concordiam.
M. D. LXXVII.
Cum Licentia Superiorum.

9   *Mechanicorum liber* (Pesaro, 1577) by Guidobaldo del Monte, patron and
scientific colleague of the young Galileo.

ics was, perhaps, the least of the subjects on the syllabus, Clavius was
encouraged to add to the Society's prestige by building up a little
school of very gifted mathematicians who would then return to their
countries to continue their specialized work.[20] It should be noted that in
1586 Clavius was far from satisfied with the state of mathematics at the

Roman College, so we should not read back into that period the reputation he had built up for his College twenty-five years later. He lamented the fact that apart from one or two, they had no one in Rome who could help the Holy See in disputes over the ecclesiastical calendar. That was a good way of bringing home to his fellow-Jesuits the fact that mathematics was not just a useless abstract subject; it was also a modest reference to his own role in preparing the revised calendar we now use, called Gregorian after the Pope who introduced it in 1582.

What we know of Galileo's visit to Clavius in 1587 is gleaned from later letters. In January 1588 he sent Clavius for criticism a theorem on centres of gravity. Clavius replied promptly to say he was out of touch with the topic but was not satisfied by Galileo's proof; a further exchange of letters left Clavius in March still reluctant to accept the proof, but he kindly advised Galileo not to treat him as an oracle (10: 22–4, 27–30). An oracular and quotable pronouncement that his work was not only original but sound was, of course, just what Galileo was looking for from such a respected expert, because he needed something of the sort to land him a post which would start him on his career. But at least he had made himself known to a leading mathematician and was being treated as a junior colleague, so the Roman trip had not been wasted.

Guidobaldo del Monte

Although the Roman visit also paid off in the form of a recommendation to Bologna in February 1588 by Cardinal Enrico Caetani, the Senate replied that they were not thinking of making an appointment at present (10: 26–7; 20: 597). That setback and whatever disappointment remained after the failure to convince Clavius of the soundness of his proof were offset in June when the proof was accepted, after initial hesitation, by no less a mathematician than Guidobaldo del Monte (1545–1607) (10: 34–5).

Galileo had been corresponding with Guidobaldo from January. It says much for Guidobaldo that from the start he accepted the unemployed and unknown twenty-four-year-old as an intellectual equal. It must have been very heartening for Galileo to find himself exchanging views on a great variety of topics with the translator and continuator of Archimedes; this was a major step towards recognition. If Clavius was kindly, Guidobaldo was more than friendly: he sent his latest book on Archimedean topics, saying he would value Galileo's opinion more than anyone else's, as it would enable him to correct copies before distributing them. Almost immediately he began to look for ways to further the career of this brilliant young man and enlisted the help of his clerical brother Francesco, who would be even more useful after he was made a cardinal in December. In July 1588 Galileo had to inform a sympathetic Guidobaldo that Fantoni had resumed the Pisan chair which he had temporarily vacated, so that particular door was shut. Guidobaldo got his brother to recommend Galileo for a grand-ducal mathematical lectureship which was being revived, again without success. The chair at Bologna went to Magini in August, but Moletti's death in March had left vacant the very desirable chair of mathematics at Padua (10: 43). Moletti, like Guidobaldo, had been impressed by Galileo's work on centres of gravity, but Galileo's plan to bring together his work on the subject and publish it was indeed confided to a delighted Guidobaldo in December but carried out only fifty years later in an appendix to the *Discourses*. Without publications a mere beginner, even one who seemed very promising to leading experts like Guidobaldo, Clavius and Moletti, could not realistically expect to land a major Italian chair of mathematics unless he first proved himself elsewhere.[21]

3

Professor at Pisa

In the academic year 1589–90 Galileo succeeded Fantoni in the mathematical chair at Pisa; his obvious abilities, his lectures on Dante and the support of Guidobaldo and his brother had been sufficient to win him the chair in his local university. His inaugural lecture was on 14 November 1589: it had been delayed because floods prevented Galileo from reaching Pisa. He still had to pay the fines attached to missing his first six lectures and further fines for an absence of eighteen days in May and June when his mother was seriously ill (19: 39, 43; 10: 43–4). Guidobaldo thought the Pisan post, with its meagre salary of 60 florins a year, was less than Galileo deserved. In April 1590 he had no news to report from Venice about the chair at Padua, but he did take the opportunity of a brief stay in Bologna to tell Bolognese friends who were unimpressed by Magini that he knew an excellent Florentine mathematician, currently lecturing at Pisa. He envied Galileo his stimulating conversations with the philosopher, Jacopo Mazzoni, who had been appointed to Pisa in 1588. But Guidobaldo was well aware that Galileo was not content with what Pisa had to offer and he kept up his efforts to find a better post (10: 42–3, 45, 47). Whatever the shortcomings of Pisa, it did provide Galileo with means of support, a chance to establish himself in the academic world and an opportunity to develop his own ideas.

The Leaning Tower Demonstration

It was during his Pisan professorship, according to Viviani, that Galileo, to the great discomfort of all the philosophers, showed by

10   A view of Pisa in *Viaggio pittorico della Toscana*, by Jacopo and Antonio Terreni, tome 2 (Florence, 1802).

observations, solid proofs and arguments that very many of Aristotle's conclusions about motion, hitherto held to be perfectly clear and indubitable, were wrong. Bodies of the same material but different weights, to take one example, fell equally quickly through air, rather than falling with speeds proportional to their weights, as Aristotle and his followers would have it. This, says Viviani, Galileo demonstrated repeatedly from the top of Pisa's Leaning Tower in the presence of colleagues, philosophers and all the students (19: 606). This anecdote has had a colourful career: it has been shunted backwards and forwards along the historiographical spectrum from a classic turning-point in human thought to a legendary piece of propaganda. It deserves more than a passing comment, as it can provide an introduction to Galileo's work on motion.

Viviani is known to have been unreliable on other points in his narrative and he was inclined to predate and exaggerate Galileo's achievements. It is highly unlikely that Galileo ever had all, or even many, of the Pisan students milling round the Campo of the cathedral: such a spectacular event would have left some trace, at least in letters if not in published works. Yet, as Drake points out, Galileo could easily have brought all *his* students and some professors to see him make a point by dropping weights from the tower.[1] Let us suppose that he did: what could he have shown them?

First of all we should forget any later ideas about showing that in a vacuum a golden guinea and a feather will fall together. Whatever else Galileo could demonstrate during his days as a Pisan lecturer it was not Newtonian ideas on free fall, as can be seen from Viviani's restriction of the demonstration to bodies of the same material.

Galileo was not the first to show that Aristotle was mistaken in thinking that the speed of falling bodies was proportional to their weight. John Philoponus had pointed this out over a thousand years earlier, as had several medieval philosophers.[2] More recently Benedetti had done the same, as had Simon Stevin in the Netherlands in 1586. Yet many Aristotelian philosophers were still content to pass on what they found in Aristotle's *Physics*. That it was not very easy to assess what the texts said by mere observation can be seen from the curious fact that before 1576 Girolamo Borro had carried out for his Pisan students a demonstration of his own, admittedly a rather rough and ready one. He took a chunk of wood and a lump of lead, which seemed to be the same weight as far as the eye could tell – there was no need to weigh them. He threw them both at once with equal force from an upper window of his house: time and again he and his students saw the lead move more slowly and hit the ground after the wood.[3]

11   Pisa Cathedral and tower: a detail from a bird's-eye view of the city in *Nouveau Théâtre d'Italie* . . . *sur les desseins de feu Monsieur Jean Bleau* [i.e. Blauel tome 1 (The Hague, 1724).

This disconcerting result makes Borro seem an unusually incompetent observer, but we should realize that this was a demonstration to convince his students by sensory experience that his theory was correct. The reported result was the one he wanted to see. The theory was that air has weight in its own place, the sphere of air: it followed, then, that because wood contains more air than lead it will fall faster through air, whereas in water wood will rise and lead will sink because air is lighter than water. Even as an intended demonstration of a truth supposed to be already known this has to be classed as rather easygoing, yet the mature Galileo would see nothing dubious in claiming to know in advance what a demonstration would show. His early writings on motion took Borro's book seriously because of the thoroughness of its treatment; they can usefully be read as a programmatic attempt to provide a preferable account of motion (1: 333–4).

Borro's demonstration may disconcert us, but to the young Galileo it was merely very defective, rather than fundamentally misconceived. On this very topic Galileo himself reported results which to us are almost as startling. It is hard to decide whether he was put to awkward shifts to explain them away or whether it was his theories which made him see things as he did. He could, it is true, do better than Borro: he could show that Aristotle was mistaken in thinking that bodies of the same material but different weights fall with speeds proportional to weight. But Galileo's first approach was to concentrate on the relative weights of the falling body and the medium, in other words, to make specific weight the key to understanding how bodies fall. This approach eventually brought him success in later controversy about bodies in water and it figures in revised form in his most mature work, but initially it also misled him. He thought, for instance, that each material had its own characteristic uniform speed of fall through a given medium such as air or water and that once this speed was attained there would be no further acceleration (1: 329). If we picture the young Pisan lecturer standing near the foot of the Leaning Tower and explaining to his students what was happening when a ball of lead fell to the ground, we have to imagine him saying that if the ball seemed to be accelerating continuously that was only because the tower was not high enough to allow it to reach its characteristic uniform speed through air, adding perhaps, as he certainly held in an unfinished Latin dialogue which I shall mention in a moment, that perspective made it difficult to judge accurately just how the ball was moving (1: 329, 406–7).

In other words, it is not impossible that Galileo dropped balls of differing weights from the Leaning Tower to show his students that heavier lead balls do not fall appreciably faster than lighter ones. But he could not have shown them anything about uniformly accelerated motion, since he thought that any accelerated motion was at most only a transient initial phase in free fall, a sort of misleading fluke, whereas Aristotle and his followers had no difficulty in accepting accelerated fall. It was also important, at this stage, to confine the demonstration to balls of the same material. From Galileo's own reports we know that in his early experiments on or observations of falling bodies he repeatedly encountered an awkward result. If he held a wooden ball in one hand and a lead ball in the other and released them simultaneously, then the wooden ball at first always outstripped the lead one, though the lead one would then overtake it and precede the wooden ball by quite a margin. Frequent tests from a high tower had satisfied Galileo that this was so (1: 334). This has all the marks of honest observation, but it was of little use to Galileo; the most that can be said for it is that it fitted in with his ideas of an impressed force which, as we shall see, moved projectiles and slowed down bodies in free fall. Cohen went to the trouble of trying to repeat the experience and did indeed find that, without realizing it, one invariably releases the lighter object slightly sooner, as can be shown by photographs.[4] What remains puzzling is Galileo's repeated observation that the heavier object eventually managed to behave properly and overtake the lighter one.

It is, therefore, tempting to assume that Galileo carried out no tests whatever, but merely meant to give a more satisfactory explanation of the phenomena Borro reported. That Galileo's most telling arguments against Borro were just that, arguments or thought experiments, will become clear in a moment. But if he was offering no more than an armchair theory, why should he claim to have made repeated tests? Nobody doubts that Borro's happy-go-lucky demonstration took place more or less as he described it. Several people realized that a high tower would provide a better testing site than an upstairs room. The Jesuit Pereira, for instance, suggested a variation to test whether falling objects speeded up because they were approaching their natural place.[5] It did not need genius to see the Leaning Tower as a convenient platform, so Galileo may well have climbed the Tower to check Borro's findings. But even a scientific genius was not likely to put a whole new science of motion on a satisfactory footing in half an hour. We need not be surprised if the young Galileo was uncritical of ideas which we find startling.

One can, then, allow to Galileo his demonstration from the Leaning Tower that Aristotle was mistaken in implying that a two-pound ball of lead would fall twice as fast as one weighing one pound, provided that one remembers that this demonstration could mark only a beginning of an independent and more acceptable account of motion. One also has to be aware that the concepts of observation or manifest experience turned out to be much trickier than they looked at first. The whole point of Borro's little demonstration, as he said in introducing it, had been to settle a theoretical dispute by appeal to 'experience, the teacher of all things'. It comes as no great surprise to find an Aristotelian professor of Greek, Giorgio Coresio, appealing to his own experiments from the Leaning Tower to vindicate Aristotle: he did this in 1612 in a controversy with Galileo himself (4: 242). What Galileo thought about that particular contribution to the controversy is, unfortunately, not known, beyond his general conclusion that he could not teach people who ignore mathematics (4: 443). So we in turn may conclude that it would have taken a lot more than any attempted demonstration from the Leaning Tower to make Aristotelians see things differently or even to give Galileo himself an adequate foundation for a science of motion.

First Accounts of Motion

I have been drawing on Galileo's first written, though unpublished, attempts to provide a satisfactory account of motion: these were brought together by Favaro in the first volume of the collected works. Most scholars follow Favaro in assigning these writings to the period of the Pisan lectureship. It has been suggested that they may belong to a later stage of Galileo's career, but I think Favaro's dating is still the likeliest.[6] In any event, there is no very obvious reason why a lecturer in mathematics should have prepared for publication a work on motion: that was usually the province of philosophers, though Galileo's own predecessor had in fact written a work on the topic. It may be, of course, that Galileo was thinking of taking up a career as a philosopher, but there is no evidence to confirm this conjecture. Since he was explicitly challenging Aristotle even in these writings on motion, it seems more likely that he already rejected the prevailing map of knowledge as reflected in the university practice of keeping mathematics and philosophy distinct. At this stage his ambitions exceeded his resources, but these writings on motion mark his first steps towards a mathematical treatment of motion. They also gave some attention to careful observation, though it would be misleading to dignify such observations with the title of experiments. If these writings are indeed from his Pisan period then they are all the more remarkable; in any case, I take them to be the earliest of his surviving treatments of motion.

The earliest one is an incomplete Latin dialogue on motion, which is set near Pisa. (Formerly many scholars took the dialogue to be later than the treatise. In an introductory sketch like this there is no great harm in lumping them together.) It opens with Dominic and Alexander taking a stroll to the seashore (1: 367–408). The dialogue reads rather woodenly, if we compare it to the mature *Dialogue on the Two Chief World Systems*, but it does show Galileo exploiting some of the possibilities of this common literary form, a form which his father had used. In particular, while Alexander plays Galileo, Dominic can raise difficulties to bring out the novel power of Alexander's ideas, while insinuating a discreet amount of flattery which would normally have had to be provided by an obliging friend in a preface or prefatory poem. Since Alexander, Dominic says, is used to the most certain, clear and subtle proofs of the divine Ptolemy and the most divine Archimedes, he can in no way assent to cruder reasoning (1: 368). Later, of course, Galileo would want to replace Ptolemy by Copernicus, but at this stage his admiration was unqualified: Ptolemy is second only to Archimedes. This is how Galileo saw himself and wanted others to see him: he would apply mathematics to the real world and replace the wordy fumblings of the philosophers with true knowledge. The dialogue was his first attempt to implement this programme. He gave a fuller and more satisfying account in a Latin treatise intended for publication, which is usually called 'The Pisan *De Motu*', a treatise which survives in manuscript, with major additions and revisions which seem to be from the same period (1: 251–366).[7]

A notable feature of this work is, again, hostility to Aristotle, who is said, for instance, to be ignorant of the elements of geometry, to use puerile arguments and to have written contrary to the truth in nearly everything he wrote on local motion. Archimedes is 'superhuman', never to be mentioned without praise, and Galileo hopes to publish soon a commentary on Ptolemy's *Almagest*.[8] In general, Galileo's method is intended to show up the shoddy method of the philosophers: he will proceed in the orderly fashion he has learned from his mathematicians, with statements depending on what has gone before and with nothing, as far as possible, taken for granted (1: 285). The forceful rhetoric may make us think that an unheard-of innovation is being announced and it is true that Galileo's later achievements go some way to justifying that impression. But a similar criticism of philosophical practice had been made more quietly by Clavius in his prolegomena to his commentary on Euclid, yet Clavius intended only to tidy up Aristotelian philosophy with minor corrections.[9] The fact that the combative style of Galileo's famous writings is already evident is worth noting. As for the content, it is more Archimedean than Aristotelian in that the long-familiar weaknesses of Aristotle's account of motion are once more exposed and an alternative is presented which makes great use of what Galileo had learned from his studies of statics and hydrostatics. Later we shall see, however, that even in the published works of his maturity Galileo's philosophy of science owed something to Aristotle and to contemporary Aristotelians among the Jesuits, while even his physics retained some Aristotelian features.

Galileo, it should be said, did not set out from the start to bring down the whole structure of Aristotelian cosmology. It was only after his discoveries with the telescope that he attempted to establish a system of the heavens to replace Aristotle's and Ptolemy's. Even here the word 'system' can mislead: in his maturity Galileo certainly tried to establish the Copernican system of the universe, but he was never a systematizer like Descartes. Yet we shall see that he also advocated and practised a reform of method: the way of understanding nature, especially motion, needed to be reformed. The application of that method by Galileo and those who followed him was to subvert much more than Aristotle's physics. The first gropings towards that method in which mathematics would be crucial were made in the early writings on motion: Galileo knew already that Aristotle's account of motion had to be replaced. It may be added that at no stage did Galileo have any thought of undermining, as distinct from revising, Catholic theology.

In dealing with natural motion upwards or downwards he first had to clarify what 'heavy' and 'light' meant; this clarification was achieved in the course of revising the treatise. Only those things should be called equally heavy (light) which are equal in volume and heaviness (lightness). We must not be misled by everyday talk of a large log of wood being heavier than a small lump of lead, because if we take equal volumes of the two materials, then the lead is certainly heavier. This clarification of heaviness and lightness implied that true heaviness, the heaviness proper to a given body, would be its weight in a vacuum. Galileo observed that the concentric arrangement of the elements of fire, water and air round the central Earth was not just a chance or arbitrary arrangement: the nearer the centre an element is, the more matter has to be squeezed into a confined space, that is, the denser the element is. This struck Galileo as very reasonable; it satisfied his conviction that nature must be well-ordered and that we can understand the reasons why it is ordered in the way it is (1: 251–3). Here, incidentally, he was drawing on ideas of the ancient atomists and he suggested that Aristotle was wrong to reject their views, a reminder that in his student days Galileo would have been introduced to Aristotle's ancient rivals. His friendly discussions with Mazzoni at this period would also have filled out whatever he had picked up about the comparative merits of Plato and Aristotle, since that was Mazzoni's speciality. In the treatise Galileo abandoned the Aristotelian idea of positive lightness and showed how different degrees of heaviness are more than sufficient (1: 289–94, 355–61).

Aristotle's opinion that bodies of the same material but different size fall through a medium with speeds proportional to their size is ridiculous, said Galileo. It is in this context that he introduced the sort of thought experiments which characterized his work from this time on. If from a tower you drop two stones, one twice the size of the other, who will ever believe that when the larger hits the ground the smaller is only halfway down? Suppose Aristotle were right: then if the two stones were joined together, they would fall more slowly than the larger stone by itself, even though the composite stone is clearly larger still and ought, if Aristotle were right, to fall faster still. Or imagine two stones equal in weight falling side by side: all agree they would fall together, so why they should fall twice as fast if they are joined to each other? No: bodies of the same material all fall together and this conclusion is not weakened by accidental oddities, such as a little sliver of lead taking longer to fall than a great lump (1: 265–6).

Nor is Aristotle any better on the role played by the medium. For him the proportion between the speeds of fall in different media is as the subtlety of the media, so that if, for instance, air is twice as subtle as water then a body will fall twice as fast through air as water. It was easy for Galileo to show that Aristotle would have no way of accounting for bodies that fall through air but float in water. He would have to live with the absurd conclusion that there is no proportion between the densities of water and air. Aristotle was simply on the wrong track. What counts, said Galileo, is relative weight, which means that it is the *difference* between the body and the medium that matters. He then showed how this works for (relatively) light and heavy bodies.

Take a piece of wood weighing 4 units and submerse it in water: let a volume of water equal to that of the wood weigh 6 units; the wood will be pushed up with a speed which is as 2. Immerse the same piece of wood in a denser medium, an equal volume of which weighs 10 units; here the wood will be pushed up with a speed which is as 6. So the speeds are as 2 : 6, not 6 : 10 as Aristotle thought. The same principle is easily applied to bodies of differing densities. So much for 'light' bodies. Heavy ones are accounted for in exactly the same way.

Take a body weighing 8 units falling through a medium of which an equal volume weighs 6 units; the speed of fall is as 2. In another less dense medium of which a volume equal to that of the wood weighs only 4 units the speed of fall will be as 4. So the speeds are as 2 : 4, not 6 : 4. Again, the principle can be applied without difficulty to bodies of differing densities.

The schematic way in which Galileo assigned these figures shows that he was merely working out an idea, an idea which was important and was undoubtedly meant to apply to the real world but was not in any sense put forward as an experimental result. In fact, he candidly admitted that his universal rules of proportion for moving bodies in the same or different media came up against a great difficulty: tests showed that his neat rules were not observed by falling bodies. A body which, according to him, ought to fall twice as fast as another from a tower did not do so. Indeed, he acknowledged the oddity mentioned earlier, namely that the lighter one starts off faster. This would need looking into – a fair indication of the limits of what Galileo could have shown by dropping weights from the Leaning Tower.

Galileo cautioned against indiscriminate talk of a body's weight. One needs to specify whether one is talking about its weight in air or water or another medium. If we could weigh it in a vacuum then, of course, we would find its true weight. Mention of the vacuum enables Galileo to point out that Aristotle's argument against the vacuum is worthless. Aristotle thought that the absence of a medium meant that a body would have to fall instantaneously, which is absurd, so there can be no vacuum. On Galileo's view it would simply fall with its proper speed, characteristic of its true weight, since there is nothing to be subtracted for resistance. Aristotle had not shown that there was any-thing absurd in bodies of different weights falling with equal speed in a vacuum. Nevertheless, Galileo pointed out explicitly that speeds of fall would vary according to their specific weights. He recognized that others such as Philoponus, Scotus and Aquinas had disagreed with Aristotle, but claimed they had not been able to refute him because they had failed to discover the principle he himself had just explained at length.[10]

Galileo was confident enough that he was improving on the at-tempts of all his predecessors, though he is noticeably silent about the ideas of Benedetti, which are often strikingly similar.[11] His confidence was not, however, mere bluff. We have seen him acknowledging awkward results that needed some kind of further attention. Nature remained similarly recalcitrant after Galileo's first attempts to use inclined planes to discover how bodies fall, a further indication that the triumphs of his maturity were not easy of achievement, although these first investigations were also presented with quiet pride as another victory for Archimedean method. The novelty which Galileo claimed for this investigation lay in the attempt to relate speed of fall to the inclination of the plane along which a body fell. His findings were plausible but not entirely convincing even to Galileo himself. But one important feature of his approach was a permanent acquisition: he took no account of accidental impediments; he stipulated that the inclined plane must be perfectly smooth and the body must be perfectly round and hard, or fluid. Later, when he had come to understand how falling bodies accelerate, he would correct his first account, but that does not alter the fact that from the beginning he was looking for simple patterns or proportions which are inevitably obscured by incidental features such as friction (1: 296–302).

The discussion of inclined planes also led him to a very important concept: a body in a horizontal plane can be moved by any force no matter how small. No matter how gentle a slope, a body which is not impeded will naturally go down it, as we see in the case of water; a body can ascend such a plane only by violent motion, that is, if it is forced up it. But if a body is in a horizontal plane, there is neither natural nor violent motion; a body in that plane will stay there and could, theoretically, be moved by any force, no matter how slight (1: 299–300). He added, however, that such a plane is not really equidistant from the horizon, since the Earth is spherical. But he did go on to consider a rotating sphere with the same centre as the Earth: it would neither approach nor recede from that centre, so its motion could not be called natural or violent; it would be neutral (1: 304–7). This was in marked contrast to the Aristotelian notion that all circular motion of the elements must be forced, since their natural motions must be towards or away from the centre of the universe. Galileo raised but did not answer the question of whether such a neutral motion would last for ever. In the dialogue he inclined to the view that it would not, since rest seems to suit earthly things better than perpetual motion (1: 373).

There has been much discussion about how close Galileo came to an understanding of inertial motion, a topic which will recur in this book. His early writings on motion do not suggest anything resembling the Newtonian model of a body remaining in its state of rest or uniform rectilinear motion unless acted upon by a force. But they are a conscious and deliberate struggling against the constraints of Aristotle's framework and they are the beginning of Galileo's own independent thought in which 'natural' and 'violent' may remain as inherited vocabulary, but lose much of their power to confine his thought to traditional channels. His comments about bodies in a plane parallel to the horizon being movable with the slightest of forces and his classification of motion about the centre as neutral are the first indications of a new approach. It is, perhaps, worth remarking that in imagining rotating bodies concentric with the Earth, Galileo did not necessarily have in mind the rotation of the Earth itself. What he wrote is compatible with a belief that the Earth rotates on its axis, but there is no other evidence that this topic was exercising his mind while he lectured at Pisa. Drake thinks that Galileo already accepted Copernicanism as a calculating device by 1590 and that this passage shows that in 1591 he was convinced that the Earth actually rotates on its axis. A few years later, in Drake's account, Galileo began to think about the cause of the tides and this led him to accept the Earth's orbital motion.[12] It is possible that this is how his ideas developed, but there is insufficient evidence to take us beyond more or less plausible conjectures.

These early writings on motion also show the influence of some important ancient and medieval criticisms of Aristotle's ideas. Such traditional criticisms made it easier for Galileo to ridicule Aristotle's explanation of the motion of projectiles. In that explanation the medium was given the role of passing on to the projectile the motion imparted by the projector. Galileo, like his predecessors, had no difficulty in showing how implausible the account was. An arrow shot into the wind, for instance, is enough to show that whatever propels the arrow it is not, as Aristotle thought, the air set in motion by the bowstring. All the ineptitudes of Aristotelianism on this topic stem from an unwillingness to accept the fact that the projectile receives an impressed force (*virtus impressa*) from the projector. Galileo's confident borrowing of this important idea from his predecessors cannot hide the fact that his explanation of projectile motion is, like theirs, entirely qualitative. To throw something up in the air is to deprive it of heaviness and make it light, as fire deprives iron of coldness by introducing heat. The lightness (the impressed force) remains in the projectile after it leaves the hand, as heat remains in iron when it is removed from the fire; both diminish gradually. So when a ball is thrown vertically, the thrower impresses on it sufficient force to overcome the body's weight. When the ball is at its highest, the impressed force has not been used up completely. It is no longer sufficient to overcome the heaviness of the body, but merely equals it; all it can do is slow down the initial stage of the natural descent of the ball to Earth. This explanation of gradual acceleration had been hit upon by Hipparchus, as Galileo discovered from Pereira, though he mistakenly thought he was being original in extending it to free fall from rest. One can see how the puzzling reports of wood initially falling faster than lead seem a little less strange on this account. All bodies in free fall will be subject to an initial retardation and will not reach their proper uniform speed of descent until the residue of the impressed force wears off. Heavier bodies are retarded more than lighter ones, so wood naturally sets off ahead of lead; but lead's proper speed is greater than wood's, so in a fall from sufficient height it can afford to give the wood a start as it is sure to overtake it before long. It should be mentioned that Galileo's treatment of projectiles is the least satisfying part of his early writings on motion. For that very reason it probably shows most clearly the difficulties facing anyone who set out to replace the Aristotelian approach (1: 308–10, 318–23, 329, 411).

Dating the Early Writings

If these writings on motion belong to Galileo's Pisan period, it is relatively easy to subscribe to the common picture of him as an independent thinker who not only saw, as others did, that Aristotelian accounts of motion were unacceptable, but also seized his first real opportunity to begin to work out a very different approach to the subject. The fact that in a biographical sketch his writings have to be described in one chapter rather than another, coupled with the fact that the documentation for his early career is scanty, whereas later it becomes overwhelming, can affect the shape that is given to his life. It is certainly more convenient to date these writings to the Pisan period: it fills in Galileo's time nicely and saves us from wondering whether a genius contented himself for three years with giving routine lectures in the University on the elements of astronomy and geometry and making enough to live on from private tuition. It also launches Galileo on his intellectual career and holds out promise of a more or less continuous development towards the mature work on motion in the *Discourses*.

So it is worth repeating that, though this convenient and intelligible account of how his intellectual career really got started may well be true, another account has something to be said for it. In this account the early writings on motion would belong, at least in part, to his Paduan period, perhaps the last years of the century or even later. A variant to be considered is that, even if the writings on motion were completed before Galileo left Pisa, it may well be that his works which Favaro classed as *juvenilia* were written at Padua in the last years of the decade, or later. These writings on scientific method, physics and cosmology can all be classed generically as Aristotelian; more specifically they show clear influence of Jesuit authors, an influence first detected by Crombie and Carugo. The researches of Wallace, on one hand, and Carugo and Crombie, on the other, may eventually lead to a consensus among scholars about their dating. There is no dispute about the significant fact that Galileo read Jesuit authors; this fact may throw light on, though it does not alter, the serious disagreements he had with some Jesuits later in his career. In any case, familiarity with an author is not proof of influence by him; the extent to which Galileo was influenced by Jesuit writers is a topic which will be mentioned occasionally in later chapters. Still, one would like to know when and how Galileo became familiar with Jesuit writings. Unfortunately the answer is not clear at present.

Wallace has published detailed research on manuscript lecture notes from the Roman College. This research gives us access to the actual teaching of Jesuits in the years around 1590. The question is: did Galileo have such access? He certainly could have done: the circulation of manuscript material of one kind or another was a significant feature of academic life, including Galileo's, and we have already seen that he had contact with Clavius. But Wallace's case rests on showing that Galileo's borrowings from Jesuit authors must have come from the lecture notes he has studied (or very similar ones). If Galileo did use such datable notes then the case for ascribing the so-called *Juvenilia*, and even the *De Motu*, to his time as a lecturer in Pisa would be pretty watertight. Yet since it is clear that he read Jesuit publications, one has to reckon with the fact that these easily accessible books may be quite sufficient to account for any obvious quotations, references or even borrowings. Crombie and Carugo are satisfied that they are sufficient. Wallace is satisfied that the massive documentation he has published from Jesuit manuscripts shows that those manuscripts must be the source of Galileo's early writings on Aristotelian philosophy, though he allows that there were other sources as well for the writings on motion.

Wallace's view seems to be acceptable to many scholars. It presents a picture of the young Pisan lecturer first mastering Aristotelian logic and scientific methodology, then tackling Aristotle's teaching on the elements and cosmology before working out his own ideas on motion, discussed in this chapter. Galileo's whole career can then be seen as a reformation of science based on Aristotle's own principles of how to discover causes, especially causes which explain what we observe. An Aristotelian understanding of the goals of science thus provides the context in which Galileo's undoubted innovations need to be understood. If Galileo came to reject Aristotelian answers wholesale, he nevertheless arrived at their replacements by methods which Aristotle would recognize as developments of ones which he himself had preached, if not practised. There is much to be said for this view (though it remains to be seen how much of an Aristotelian Galileo was). Carugo and Crombie have quite independently established a large part of same picture, but they would account for Jesuit-Aristotelian ideas in Galileo's works without any appeal to the Jesuit handwritten lecture notes which are the basis of Wallace's dating. So, if we think of Galileo's life as a film, Crombie and Carugo are insisting that we cannot at present be sure of the correct order of the reels.

It must also be repeated that the early writings on motion make excellent sense if read as Galileo's answer to Borro's book, supplemented by Galileo's reading of well-known published works. Nor is there any reason to play down his own insistence that his aim was to follow the method of mathematicians, especially Archimedes. It took him almost another fifty years to publish a new science of motion, but a serious beginning was made in the fragmentary dialogue and the treatise. In the present state of scholarship it seems reasonable to assign these writings to the Pisan period, as I have done in this chapter. But this dating does not, in fact, depend on establishing that Galileo used Jesuit lecture notes at this period.[13]

New Responsibilities

One thing that is known for certain about this period had a great influence on the rest of Galileo's life. His father died at the end of June or on 1 July 1591; on 2 July he was buried in Santa Croce in accordance with ancestral tradition. All Galileo's worries about his meagre salary were greatly increased, as he was now head of his family. It was no longer a question of promising a well-chosen present for his sister's wedding (10: 46). He had to take on the serious burden of providing her dowry (10: 61). Galileo was never a family man in the usual domestic sense: his own children were all born out of wedlock and he never married. But he had, nevertheless, a real sense of his duty to provide for his near relatives: to him it was all part of what was expected of a Florentine gentleman. He was also a gregarious and generous man himself and not averse to living beyond his means, so much of his scientific work over the next two decades was punctuated by negotiations for advances of salary and increases in salary, supplemented by occasional borrowing. A good deal of his financial worries were caused by the misfortunes or shiftlessness of those for whom he considered himself responsible. His father's death must have brought home to him the poorness of his prospects at Pisa (10: 47). It may be that, even without that new incentive, he would have had to look for employment elsewhere, since it was not by any means certain that his Pisan contract would be renewed when it ran out in 1592.

Academic Dress

That the young Pisan professor was something of a nonconformist by temperament appears clearly in some jaunty and earthy verses of his against wearing academic dress (9: 213–32). He had in fact been fined at the end of his first year for not being properly attired (10: 44). These 300 lines trace the world's evils to the abandoning of the original nakedness of Paradise. All the distinctions of degree which separate human beings from each other and allow some to lord it over others are tied in with differences of costume. Not that the young Galileo was a leveller who found it unnatural to conform to authority or to defer to the privileges of rank. Throughout his life he was alert to what was due to his own honour, position and rank and quite an assiduous courtier; so far from being politically subversive, even in thought, he would count as politically naïve compared with friends he later made in Venice, such as Sarpi or Sagredo. Yet we have seen him attacking the established Aristotelian philosophy in a work which was certainly intended for publication, so some private fun at the expense of his professorial colleagues colours the picture nicely. In his description of them bustling about in their long academic dress or togas, surrounded by a crowd of admiring students, we catch a glimpse of what Galileo was like in unbuttoned mood, making routine fun of priests and friars and exposing the customary paraphernalia of academic success as irritating and petty humbug. The brief sketches he gives us of everyday life at the University let us see that he lived in an atmosphere which was not quite as monastic or seminaristic as the official regulations might lead us to think. The verses were, of course, just for his friends to see, but they reinforce the impression that he would not be content for long with what the University of Pisa had to offer. There is no reason to think that many of his professorial colleagues would have missed him. In any case, he would be much better off in all respects at Padua, if only he could get there.

Fortunately for Galileo, Guidobaldo was a conscientious patron, he did have a brother who was a cardinal and he was well qualified to recommend people to mathematical posts. So it was probably a matter for relief rather than surprise when on 26 September 1592 Galileo was appointed professor of mathematics at the University of Padua for a period of four years, renewable thereafter if he gave satisfaction (19:

111–12). Guidobaldo knew how to handle these things: he disclaimed all credit, attributing the appointment entirely to Galileo's incontestable merit, a courteous and encouraging touch from an established expert to a promising young lecturer who might have languished longer in uncongenial surroundings without the help of an influential and discerning friend (10: 54).

4

The Proper Home for his Ability

Galileo's application for the chair at Padua had had the support not only of Guidobaldo but also of a person who was the centre of cultural life in Padua, Giovanni Pinelli. This Genoese nobleman was born in Naples but in 1558 he had settled in Padua. His house, the meeting-place for all the liveliest minds in Padua, was where Galileo lived until he found lodgings of his own. Pinelli had a good collection of scientific instruments but was best known for his vast library, which he made available to scholars. Much of the library was lost at sea after his death in 1601, but the residue included manuscript works by Galileo (10: 47–8, 51). To the gregarious Galileo, Pinelli's and other intellectual circles were a cherished feature of Paduan and Venetian life.

Despite the high reputation of the University, the professor of mathematics was not well paid, since his was not a principal subject like philosophy. He was expected to supplement his salary of 180 florins by private tuition and even to take in boarders. Galileo did both and later set up a small workshop in his house to manufacture instruments, so that he presided over a lively little household of students of mathematical subjects such as military architecture, elementary astronomy and perspective. The stimulating conviviality of the household was a happy memory for many a foreign pupil. Now that Galileo was head of his family his financial needs were pressing. For many years he was hard put to find the money to help his brother to establish himself as a musician and to pay the over-generous dowries he guaranteed his sisters: he was already responsible for Virginia's and within a decade

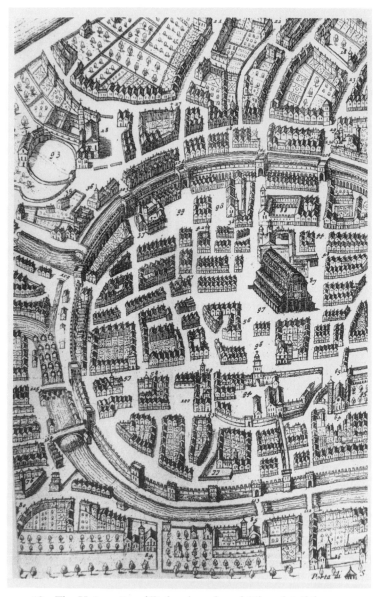

12   The University of Padua (numbered 88): a detail from a
bird's-eye view of the city in *Nouveau Théâtre d'Italie . . . sur les desseins
de feu Monsieur Jean Bleau* [i.e. Blaue], tome 1 (The Hague,
1724).

13   Girolamo Mercuriale, friend of Galileo, from Isaac Bullart, *Académie des Sciences et des Arts*, tome 2, livre 2 (Amsterdam, 1682).

had to provide for Livia. So far from getting the help he had optimistically counted on from Michelangiolo, he was to find himself frequently having to subsidize his brother as well. Still, he was where he had long wanted to be. Girolamo Mercuriale, who had just accepted a chair in

Pisa to round off a distinguished career as professor of medicine, knew Padua well from his days as a professor there: he assured Galileo that it was the proper home for his ability (10: 54). Galileo made a fine start in this new home with a splendid inaugural lecture on 7 December 1592 (10: 51–3). His lecture course began on 13 December (19: 21).

An indication of both the new challenges Padua and Venice offered and the important contacts Pinelli could provide came in March 1593, when Galileo was consulted by Giacomo Contarini on behalf of Venice's world-famous Arsenal about the most efficient way of arranging ships' oars (10: 52, 55–60). Galileo later paid tribute to the skills of the Venetian shipwrights in the very first words of his *Discourses*: frequent visits to the Arsenal open up to speculative minds a large field for philosophizing, especially in mechanics, since the workmen, what with their predecessors' observations and the ones they keep adding themselves, must be completely expert and knowledgeable. Quite right, confirms the next speaker, who may also be taken to speak for Galileo: the foremen, particularly, can often help to explain things which seem not just wonderful but almost incredible (8: 49). Galileo was good with his hands, in contrast with Kepler, who described himself as an unhandy theoretician (10: 507), and he was fascinated by how things work. This was not the literary curiosity which one can find in plenty of Aristotelians, who were happy to enliven their pages by uncritical reports of astounding things. It was the penetrating involvement of a man who wanted to understand the world. Aristotelian philosophers were fond of repeating their master's dictum about nature being impossible to understand without understanding change, especially motion. But the typical Aristotelian philosopher, if admitted to the Arsenal, would have been a mere tourist: it would not have occurred to him that what he was seeing had anything important to tell him about the philosophy of nature, about physics, about the science of motion.

Too much should not be read into this reference to the Arsenal or to Contarini's consultation. Galileo was in no way a consultant engineer to the Republic, though he did advise the Grand Dukes of Tuscany on more than one occasion. But (at least if we confine ourselves to datable evidence) his early years at Padua were largely given over to study of practical topics, much nearer to the world of shipwrights and soldiers than that of the philosophers. In February 1594 he was granted a patent in Venice for a device which would raise water for irrigation using only one horse (19: 126–9). It is not known what made this machine sufficiently different from others to justify a patent, but Niccolò Contarini paid the expenses for a working model in 1601 (19: 202), a model which in April 1602 Galileo promised to bring to Florence (10: 87–8). More than thirty years later one of his former pupils and supportive foreign admirers, Nicolas Fabri de Peiresc, still remembered seeing Galileo demonstrate the model at that time (16: 27). The fact that Galileo could work on such a machine is a fair indication that his energies were by no means exhausted by his teaching obligations and that he had a personal interest in technology.

His practical turn of mind is also very evident in the courses he gave to private students on military architecture and fortification. He may even have given his first professorial course on this subject, since he was allowed to choose his own topic that year. This may be considered a relatively new applied science, which Italians had had to learn the hard way in the early years of the century. The great new factor was the use of effective artillery: several mathematicians had realized that understanding the path of projectiles was essential to a truly effective use of cannon. Galileo later claimed that what led him to study motion in the first place was a desire to discover the path of projectiles, but he had not made much progress at Pisa (14: 386). In fact, his private lectures were more concerned with explaining the best means of defending positions by fortification and the best ways of overcoming those defences. He was clearly intrigued by such problems and expounded them with great clarity (2: 15–146). There is little trace of the revulsion at the horrors of contemporary warfare which was so pronounced in his favourite poet, Ariosto.[1] Perhaps the apolitical Galileo did not give the matter much thought, at a time when Italy was relatively peaceful; as far as he was concerned, he was just giving the private courses of applied mathematics expected of someone in his position. In connection with these courses be began, from 1598, to make and improve an instrument which he called 'the geometrical and military compass', the subject of controversy later on.

Cosmography

One of Galileo's regular professorial tasks was to give an elementary course in cosmography, an exposition of the rudiments of astronomy and geography, which was part of a general education and would be of special interest to medical students, who would need it if they were to make use of astrology in their professional work. Favaro published an Italian text from a manuscript copied in 1605 (2: 211–55): this would have been for use with Galileo's private pupils, but it is reasonable to assume that his Latin lectures in the University (and at Pisa, for that matter) would have been very similar. As one would expect, the exposition is clear: only a very incompetent lecturer would have had difficulty in producing clear lectures on such a well-worn topic. We know from other sources that Galileo did not espouse Copernicanism in his public lectures, so there is not much to be deduced from his incidental references to heliocentrism.

But he did not say anything hostile: whether the Earth moves is a question worthy of consideration, given that some great philosophers and mathematicians have thought it a mobile star. Nevertheless, he said, since we are following the opinion of Aristotle and Ptolemy, we shall bring forward those reasons which can lead one to believe that the Earth is quite stable. It is easy to show it does not move in a straight line. But that it can move circularly is something which has more verisimilitude and so has been believed by some, especially since it seems to them almost impossible that the whole universe apart from the Earth should revolve in twenty-four hours; they would prefer to have the Earth revolve in the opposite direction in the same period. Galileo then rehearses the arguments used by Ptolemy to refute the opinion and leaves the matter there (2: 223–4). Now, as we shall see in a moment, he counted himself a Copernican in 1597. The incidental remarks in the 1605 cosmography are compatible with his having temporarily lost interest in Copernicanism; they are also compatible with semi-Copernicanism, a belief in the Earth's rotation on its axis without any annual revolution around the Sun; finally, they are compatible with a continued conviction that Copernicus had discovered the true system of the universe. He may have felt an obligation to expound geocentrism in his public lectures and need not have felt any call to discuss heliocentrism privately. In fact, these lectures tell us nothing definite about how his ideas developed. If we had all his earlier courses on cosmography, whether private or public, they might be virtually identical with this one.

Copernicanism

The first definite information about Galileo's Copernicanism comes in a very friendly letter he wrote on 30 May 1597 to Jacopo Mazzoni, who had just published his book comparing Plato and Aristotle (2: 197–202). Galileo was pleased to see that Mazzoni had become more inclined to accept ideas which Galileo had espoused in the conversations they used to have at Pisa. This may well refer to the ideas on motion which we saw in the preceding chapter. Galileo certainly takes it that they both acknowledge Plato as master, as dedicated investigators of the truth should, a clear enough reference to Plato's endorsement of mathematics as essential in the study of nature. But he was at first dismayed to see how much opposed Mazzoni was to Copernicanism, since Galileo thought that opinion considerably more probable than Aristotle's and Ptolemy's. (It may be significant that he does not write as though Mazzoni already knows him to be a Copernican.)

One great difficulty in accepting Copernicanism was that no one could observe any change in the aspect of the fixed stars. Everyone had long been content to accept that the distance to the stars is so great that the radius of the Earth is negligible in comparison, so when we see the stars from the Earth's surface it is as though we are viewing them from the centre of the Earth itself. In other words, every educated person accepted that the size of the Earth is negligible in comparison with the vastness of the universe. But anyone who claimed, as Copernicans did, that the Earth circles the Sun in an orbit with a radius of perhaps 5 million miles (using the traditional Earth-Sun distance of about 1,200 terrestrial radii), had to concede that the stars still looked the same. This meant that they are so far away that the huge radius of the Earth's orbit is negligible and it is still as though we were at the centre of the universe. In other words, if the Earth moves in a large orbit, the already vast universe has to be enormously inflated, filled out with an unimaginable volume of waste space, and all to let Copernicans say that the Earth is a planet. The supposed size of the Copernican solar system and universe may seem rather cosy nowadays, but it so staggered some able minds that to them it seemed sufficient to show that Copernicanism was false. Even for Copernicans the fact that stars looked the same even when supposedly viewed from points separated by 10 million miles, that is, the inability to detect annual stellar parallax, was a very real difficulty, one which they could (in others' opinion) only explain away. The difficulty was not to be resolved in their favour by reliable observation until the 1830s, by which time astronomers knew that the distance from Earth to Sun is about 93 million miles. Galileo, of course, could not point to any change in the aspect of the stars that would reflect the Earth's annual motion, though he thought such parallax would eventually be detectable; but he could and did show that the particular argument adduced by Mazzoni was based on a misunderstanding. In other words, he went to some trouble to show Mazzoni that Copernicanism was not so easily disposed of. The size of the universe did not worry him: it would be as big as research showed and could not be limited by our inherited parochialism.

On 4 August 1597 Galileo told no less a person than Kepler that he was a Copernican. Kepler had given a friend two copies of his recently published *Mysterium Cosmographicum* to take to Italy, presumably to leave them where they would do most good. Galileo, whom Kepler had never heard of, was the lucky recipient; he wrote an acknowledgement within hours of receiving the book, after reading only the preface. That was enough, he wrote, to make him glad that he had found such a companion in searching out the truth; it was a sorry thing that those who cared for the truth and did not follow a perverse method of philosophizing were quite hard to find. So he congratulated Kepler on his beautiful discoveries confirming the truth and promised to read the book: he would do this the more readily since he had come over to Copernicus's opinion many years ago and on its basis had discovered the causes of many natural effects which could not be accounted for on the common hypothesis of a fixed Earth. He had written many reasons in favour of Copernicanism and solved many of the objections to it, but he did not dare to publish them because he was frightened by the fate of Copernicus, who gained immortal fame with a few, but ridicule from fools, whose number is infinite. He would, indeed, publish if there were more people like Kepler, but as things were, he would give the matter over. In the meantime, letters from Kepler would be very welcome (10: 68–9).

That Galileo had been a Copernican for *many* years is at least as doubtful as his claim that he had discovered many effects explicable only by the Earth's motions. He could be meticulous in scientific measurement when he chose to be but was prone to breezy exaggeration in everyday life, and even in published work. What may surprise some is the timidity evinced in this letter, since he is usually portrayed as a bull in a china shop. Yet an autobiographical aside in the later *Dialogue* tells us quite plainly that without the telescopic discoveries Galileo would probably have remained a good deal more reluctant to accept the Copernican system. Those discoveries certainly decided him to defend Copernicanism publicly (7: 356). Timid or not, the letter was good news for Kepler, who replied promptly enough, sending another two copies of his book, encouraging Galileo to take heart about the prospects of Copernicanism and asking him at least to communicate his findings privately (10: 69–71). Galileo did not reply. One has to conclude that, despite their common heliocentrism and passionate interest in mathematics, they were too dissimilar in outlook and method for Galileo to be comfortable with Kepler. Galileo had no time for the heady and mystical Pythagoreanism which carried Kepler to his greatest achievements, which came only a dozen years later. In any event, such cooperation as there was later between the two great scientists was only possible because of Kepler's engaging readiness to oblige someone who kept his distance when he chose. On the other hand, on this first contact between them, it is not surprising that Galileo did not respond to Kepler's suggestion that he should try to detect annual parallactic motion in the fixed stars. If Galileo could have done that, he would surely have told all Europe, since it was the one thing, as they both knew, that could have convinced astronomers that Copernicanism was the true system. Kepler had to content himself with the shrewd guess that Galileo must think the tides are caused by the Earth's motions, whereas he thought they must be caused by the Moon (10: 72). That the tides can only be accounted for by the Earth's motions was to become a favourite argument of Galileo's; he seems to have hit on it in 1595. It provided an ingenious but mistaken climax to his *Dialogue*.[2]

Salary

By 1598 Galileo had to think of renewing his contract with the University. There was not likely to be any difficulty in the renewal itself: the problem would be to obtain a significant increase in salary. Galileo turned for help to Giovanfrancesco Sagredo (1571–1620). Sagredo, a well-connected patrician, was willing enough to practise the art of the possible to help a friend. He became, in fact, Galileo's closest friend, fittingly commemorated as a participant in Galileo's *Dialogue* and *Discourses*. His later letters to Galileo show a man at ease with the world, cheerfully detached in his sardonic judgements of politics, well able to enjoy life and genuinely interested in science, where he was no mere dabbler. Sagredo could put Galileo's case directly to the decision-makers, the three *Riformatori* of the University. He sent Galileo a verbatim account of how the negotiation concluded in September 1599. The upshot was indeed a renewed contract, with an increase in salary from the original 180 florins to 320, but the courteous fencing of one of the *Riformatori*, Leonardo Donato, and the blank disdain of another, Zaccaria Contarini, showed that Sagredo and Galileo were thought to be too pushy. Donato pointed out that Galileo's predecessor, Moletti, never got more than 300 florins – the implication being that there was no reason to rate Galileo's merits higher. What Bologna paid (to Magini) was irrelevant: money was short in Padua. The salary was never meant to provide a living: it was to be supplemented by fees from private teaching. If Sagredo pressed for more than 350 florins he would be asking for something that would throw the University into turmoil, hardly suitable conduct for a Venetian gentleman who had already done enough to satisfy his obligations of friendship to Galileo. So let the matter rest there and let Galileo be content with what was available. Sagredo kept pressing, but Donato was courteously immovable. Contarini, bored by the whole business, was exasperated with Sagredo and astonished at Galileo's pretensions. He had already made clear that they could easily look elsewhere for someone (10: 77–8).

This frank little report throws a clearer light on Galileo than many of the jibes of his opponents or encomiums of his friends. Almost halfway through his life and after a decade of public teaching he was considered a worthy incumbent of a minor academic post, good enough to be given the sort of salary that a comparable predecessor achieved only towards the end of his career, good enough, in other words, to be accorded sufficient encouragement to keep him at Padua, but not so outstanding that he could not be replaced if he insisted on a salary that would cause widespread comment. Sagredo can be given credit for realizing that Galileo was worth more. The *Riformatori*, who had to go by public performance and achievements, seem to have made a sensible enough award.

Family

Part of Galileo's recurrent financial difficulties was caused by the obligations he assumed to provide his sisters' dowries. Michelangiolo was a burden rather than a help, but he did make a good point when he complained that Galileo should not have committed them both to dowries which were beyond their means: was he expected, he wrote from Munich in 1608, to have postponed his own marriage in order to support his sisters (10: 192–3)? That was a bit tactless, since one expense Galileo never incurred was a wedding feast for a wife of his own. For a decade he lived in a stable relationship with Marina Gamba, a Venetian. She bore him three children: Virginia in August 1600; Livia in August 1601; and Vincenzio in August 1606 (19: 15–16). Marina was not envisaged in Galileo's later plan for a life of research and publication in Florence: she and Galileo seem to have separated quite amicably, or perhaps in a businesslike way, when Galileo left Padua in 1610. Galileo was a conscientious enough father in providing for Vincenzio's education and eventually having him legitimized. His treatment of his daughters seems more offhand. He sought to place them in a convent as soon as possible, sooner in fact than was actually allowed by Church law: Virginia was only nine years old when he made his first soundings (10: 306). Not even Cardinal del Monte could get round the law, when

Galileo approached him eighteen months later: the girls could not be accepted until they were old enough, not even as mere lodgers who would wait to take the veil at sixteen (11: 234, 245). It was to be October 1613 before Galileo managed to place them in the convent of San Matteo in Arcetri near Florence, this time with the help of Cardinal Bandini, when to later ways of thinking neither girl was much more than a child (11: 588). In Virginia's case this turned out well: busy and happy, she became a great support to Galileo for a decade or more until her early death in 1634 at the most troubled time of his life. Her lively and affectionate letters give us the only real glimpse of anything that we can call Galileo's own family life. As Sister Maria Celeste she provided something unique in Galileo's eventful life and she has won the hearts of even his anticlerical devotees. Livia, by contrast, was chronically unhappy as Sister Arcangiola: if she was not as affectionate towards her father, he could hardly blame her. Whether he ever felt he should blame himself is impossible to say. According to the customs of the time and place he had provided adequately for all his children.

Even his inveterate opponents in later controversy did not suggest that by his irregular domestic arrangements Galileo had forfeited his claim to be counted as a good Catholic. They conceded that he was orthodox in intent, though misguided and too inclined to correspond with Protestants and other dubious characters. 'Good Catholic' did not mean 'exemplary Catholic': it just meant 'definitely one of us', a fact which his many clerical friends, including cardinals, never had any reason to doubt. The fathering of children out of wedlock would not have closed many doors to him in Venice or Padua. Florentines, however, would probably have judged it imprudent to bring Marina to the Tuscan court, even if she had wanted to come. Most biographers of Galileo have simply recorded the bare facts, accepting that to say anything more about his family life in Padua is mere speculation; that seems very sensible.

How Marina and her children fitted into the lively Paduan household, with its changing population of foreign boarders and daily pupils, is not known. At least one of the pupils, the most able and the best known to posterity, did not lose touch with the family; in fact, he almost became part of it and gave Galileo enduring support, which would have been remarkable even in a son. This was Benedetto Castelli (1578–1643), a Benedictine monk, who first studied under Galileo both at the University and privately from about 1604. Within a few years he was collaborating with Galileo. Later, when he had made his own reputation as a professor of mathematics and an unrivalled expert on all problems connected with the flow and control of water, he liked to consult Galileo, whose own expertise in this area was considerable. To the very end he did all he could to further Galileo's discoveries and to support him in his misfortunes.

The little we know of Galileo's own family life at Padua might give the impression that he was cold or self-centred: the wide range of his many friendships and the unwavering devotion of several close friends such as Castelli or Sagredo show that, whatever his faults, he was not cold. The help he gave to Castelli himself, or later to Castelli's brilliant pupil, the mathematician Bonaventura Cavalieri, or to dozens of others we know about (including an impoverished servant), does not fit in with the picture of a self-centred man. He was, however, ambitious for lasting intellectual glory. In pursuing that perfectly honourable ambition he was sometimes unfair to opponents in controversy. We shall see evidence enough of that soon, but for all we know, his undoubtedly happy years at Padua were also happy for all his household, children and mistress included.

Motion

Galileo's early writings on motion set out to provide a comprehensive alternative to the unsatisfactory accounts still taught by Aristotelian philosophers such as Borro. In those early gropings towards a mathematical description of how bodies move, he may not have been clear in his own mind about how closely such mathematical descriptions must fit the observed facts if they were to count as satisfactory, but he had already realized that the ways in which bodies really move could be understood only by concentrating on selected aspects, while disregarding others as incidental features to be accounted for separately. This was enough to mark him off from a typical philosopher. To tell an Aristotelian not to bother about friction or air resistance was like telling him to abandon the real world we all inhabit in favour of a mathematical fantasy. Yet Galileo already had a pretty good idea that this was the only way forward if he was to do for dynamics what Archimedes had done for statics. He was still Aristotelian enough to see his own work as a search for causes: he hoped to reach the true, unique causes which explain the motions we see. But somehow he taught himself the lesson that the first thing to do was to understand properly the effects whose causes were to be sought – in itself a lesson that was at least compatible with Aristotle's philosophy of science. In fact, his ultimate achievements consisted in providing a method for describing mathematically how motions take place (in kinematics, as we would say), rather than in giving a new array of causes such as forces to replace Aristotelian or medieval explanations of motion (that is, in dynamics).

In chapter 9 we shall see something of Galileo's own presentation of his greatest achievements in his *Discourses* of 1638. That presentation will make evident how far he had progressed from his first writings on motion. But it will also raise the question: how do those results relate to the real world and to the way he actually discovered them? This question has puzzled many since the time of publication. That some of his mature reports of experiments or demonstrations seem careless or mistaken is perhaps a minor problem, but that some seem too good to be true is more disconcerting and has led able scholars to class them as imaginary. Yet the scholarship of the last three decades has done much to rehabilitate Galileo's reputation as an experimentalist, while still allowing that *some* of his reported experiments are idealized or imaginary. This is still a controversial area, but it has long been known that Galileo made discoveries of lasting importance during his Paduan period from about 1602 onwards. There is now little doubt that these discoveries involved serious practical work of measurement: at the very least such measurement must have served to make his developing theories evident and to that extent to confirm them. It is, however, a distinct question whether such experimental work was the actual path to discovery. Even if it was, the traces of such work in undated papers leave room for much debate among researchers about the actual route he took or what to conclude from careful reconstructions of the experiments or demonstrations he made. This at least enables the rest of us to sympathize with seventeenth-century Aristotelian philosophers, who were in the similar position of trying to assess from books what demonstrations like Galileo's showed, though, unlike us, they had nowhere else to turn to for the 'right answers'.

We have already seen his first tentative approaches towards a crucial 'right answer' in what are sometimes called his 'proto-inertial' ideas: we have seen him beginning to question the Aristotelian model, where rest is self-explanatory and only motion needs explanation. Although he never got quite as far as the Newtonian model, in which a body will continue in its state of rest or uniform motion in a straight line unless interfered with, he is rightly credited with a restricted form of the law of inertia, which was an adequate basis for the discoveries presented in the *Discourses*. Two of these discoveries belong to the latter part of his Paduan period: the law of freely falling bodies and the parabolic motion of projectiles. A letter to Guidobaldo shows us how far he had got in November 1602 (10: 97–100).

Galileo was trying to convince Guidobaldo of a strange and wonderful fact: all motions from any point in a vertical quarter-circle will take place in equal times. Guidobaldo remained unconvinced after he had rolled balls in the rim of a sieve, but this crude test was too exposed to disturbing factors to put Galileo off. He proposed to Guidobaldo a much neater way of showing how motion really takes place. Attach two balls to threads of equal length (about four to six feet), hang the threads from nails and pull one back a good way and the other a short distance from the perpendicular: when released simultaneously they swing through unequal arcs but do not get out of step with each other even in a hundred swings. So, since the fall through the large arc clearly takes the same time as that through the small one, it is obvious that all falls through any arc of a quarter-circle will take an equal time, so long as one disregards any aberration caused by things like inequalities in the surface or curvature of an actual circle such as the rim of a sieve. If this seems hard to accept, Galileo goes on, he has discovered something even more remarkable: the times of descent are equal along all chords drawn from the highest or lowest points of a vertical circle.[3]

One of Galileo's main interests in his early years at Padua had been the study of mechanics. Any full account of the way his ideas developed would have to devote a chapter to that topic, but an introductory sketch can at least note how, by 1602, he had not only moved away from Aristotelian interests but had entered a new area where even a Guidobaldo (and Galileo himself) could expect to be surprised at what applied mathematics would show. Later we shall have to look more closely at what he said about pendulums and motions on inclined planes. But we need not imagine that at this stage Galileo could read off, as it were, all the things which he would be able to demonstrate later from, for instance, the motion of pendulums. When he wrote to Guidobaldo he may still have had no clear ideas on the acceleration of falling bodies. In any case, he conceded rather ruefully that the abstract science of geometry seemed to lose its certainly when applied to the material world, with all its accidental properties.

It was probably in 1603 that Galileo contracted the rheumatic illness that was to disrupt the rest of his life with incapacitating seizures. But his study of motion was only interrupted: he soon made great progress, as we know from a famous letter which he wrote to Paolo Sarpi on 16 October 1604 (10: 115–16). Sarpi (1552–1623) is generally remembered for two things: first, his role as Venice's counsellor in a bitter jurisdictional dispute with the Pope, which came to a head in 1606 and took a year to patch up, and second, his brilliant, if unreliable, *History of the Council of Trent*.[4] But his versatility extended to mathematics. Three years later Galileo was to say 'without any hyperbole' that no one in

Europe surpassed Sarpi in knowledge of mathematical sciences (2: 549). That, however, was Galileo at his most rhetorical, in full cry after the hapless plagiarist of his compass, so a little brazen hyperbole was a help towards showing that condemnation by Sarpi left the victim with no appeal. Still, Sarpi was certainly at least a very gifted amateur and good enough for Galileo to bounce important ideas off him.

This letter shows that Galileo is now committed to the deductive mathematical method of exposition which characterizes the third and fourth days of the *Discourses*. Admittedly he reports that he has not been successful in his search for an indubitable principle, an axiom from which he can demonstrate those properties of motion that he has discovered. He has had to make do with a principle which only has 'much of the natural and evident' about it: with this principle he can demonstrate that the spaces (distances) passed in natural motion (free fall) are proportional to the squares of the times and that the spaces passed in equal times are as the odd numbers starting from unity. We shall come back to the principle in a moment, after looking first at what may be called Galileo's law of natural motion (free fall). If the successive odd numbers starting from unity represent the spaces passed in successive equal times, one only has to write out the series 1, 3, 5, 7, 9 . . . to see that the total space traversed after the first interval is 1, after the second 4, after the third 9, after the fourth 16 and so on; so the spaces passed in free fall from rest are as the squares of the times. So Galileo had not only come to see that bodies in free fall accelerate; he recognized that acceleration as uniform and could already describe it in a beautifully simple mathematical law. This times-squared law was a permanent acquisition and was presented with justified pride in the *Discourses* as one of Galileo's great discoveries.

Yet the point of the letter to Sarpi was not to announce this as a discovery – he must have been told about it earlier – but to announce a principle from which this law followed. The principle was that the increase of speed of a falling body is proportional to the distance from its starting point. By 1609 Galileo had rejected this principle as mistaken, but he still thought the mistake sufficiently natural and instructive to be worth mentioning in the *Discourses* as evidence of how difficult it was to arrive at a true law of nature (in this case, that the speed of fall is proportional to the *time* elapsed from rest and so to the square root of the distance).[5]

It may well be that it was work on the path of projectiles that led him to abandon the mistaken principle. By 1600 Guidobaldo had performed a rough experiment in this connection: he threw an inked ball onto a steeply sloped plane; the splotchy path traced by the ball seemed symmetrical, like a rope suspended from a nail at each end, though inverted in this case, since the ball first went up, then down. This catenary curve, as Guidobaldo remarked, was similar in appearance to the parabola or hyperbola.[6] Now from his work on mechanics Galileo probably already understood that horizontal and vertical motions are independent of each other; this was a crucial break from Aristotelianism, where there could be no mixing of motions without one interfering with the other. He had stated that a body at rest in a horizontal plane could be moved by any force however slight; he realized that once set in motion such a body would preserve its motion – another idea completely at odds with Aristotelianism. His law of free fall, where distance is proportional to the time squared, would give him the independent vertical motion of a projectile.[7] So it may well be, as Naylor has proposed, that an undated working paper (perhaps from 1605) actually shows Galileo setting out to investigate the trajectory of a projectile, guided by the idea that it could be parabolic.[8] He did this by rolling a ball down slopes of different inclines, set near the edge of a table: for each incline he measured the trajectory through the air and found it was a semi-parabola. Other undated papers show similar experiments, work which was to be dressed up for publication in the *Discourses*. When Galileo claimed in 1632 that the thing which had first provoked him to study motion more than forty years earlier was to discover the line taken by projectiles, his point was that once he had hit on the correct line then it was relatively easy to prove that it was the correct one (14: 386).

The preceding paragraph gives the merest glimpse of the sort of important work done by Naylor, Drake, Wisan, Hill and others in the last two decades on a topic crucial to understanding how Galileo's thought developed. There is as yet no consensus on how he arrived at his most important discoveries, since the proper interpretation of crucial working papers (including the ones alluded to above) is still disputed.[9] But there is no doubt that at this period Galileo did perform experiments with pendulums and inclined planes; this empirical work seems to be an integral part of his method of discovery. That may not seem a very resounding conclusion, since such experimentation is the core of the traditional account of Galileo's work commonly given a brief parade even in textbooks of physics. But it does correct an influential view, based on Galileo's published works and sustained with great erudition by Alexandre Koyré, that Galileo's achievements were mathematical and theoretical while his alleged experiments were largely imaginary.[10] Even a glimpse of one reconstruction of his experimenting with falling bodies is enough to show how far his work at Padua had brought him from the already innovative early work on motion. It makes it less surprising that before he left Padua he had a solid basis for his new science of motion.

The New Star

In October 1604 a great stir was caused by the appearance of a 'new star'. It immediately reminded people of the similar phenomenon of 1572, which Galileo had been shown as a child. That star is known as 'Tycho's nova', because the Danish astronomer observed it with great care and established that it was sited among the stars; the 1604 star came to be known as 'Kepler's nova'. Both are now classed as supernovae, exploding stars of exceptional brightness; these were so bright that they could be detected with the naked eye – as yet, there were no telescopes. It is a curiosity that two such rare occurrences were seen during the period when there was so much discussion about the prevailing Aristotelian world-picture. It is another kind of curiosity that both together were not sufficient to make determined followers of Aristotle change their opinion about the nature of the heavens. Despite Tycho's careful work, the great question for many people in 1604 was still whether the startling phenomenon was a heavenly body at all, let alone a star.

The first sighting at Padua, and one of the first anywhere, was by Simon Mayr and his pupil, Baldassare Capra, on 10 October, as they soon informed Galileo through a common friend, Giacomo Cornaro. Presumably this was after Galileo's letter of the 16th to Sarpi, one of several instances where Galileo's study of motion was interrupted by unpredictable diversions. Capra later complained in print that Galileo did not mention Cornaro in his lectures on the topic; worse than that, Capra definitely left his readers with the impression that Galileo had not given them proper credit for priority of discovery (2: 293–4). Capra also made various criticisms of Galileo's lectures, which would probably have remained unnoticed if, in 1607, he had not engaged in the spectacular folly of publishing on Galileo's own doorstep a work in which he claimed Galileo's geometrical compass as his own. In Galileo's devastating published demolition of the foolish Capra the first item was putting the record straight about the new star. There Galileo said that in his first lecture he had given full and handsome credit to the Paduan discoverers of the new star, and no mention of Cornaro was called for. He also repeated what he had jotted down at the time in a testy note: that it was no disgrace for the university mathematician to have missed the first appearance of the star, as though he was obliged to keep watch every night of his life to see whether a new star might appear (2: 278–9, 520–1). He went on to show Capra's ignorance of astronomy, which was lamentable in someone publishing on the subject and culpable in one attributing to Galileo views which Galileo had already kindly told him through a friend that he did not hold. So much for Capra, who was left with little credit as an astronomer, except his share of having been with Mayr the first to see the new star in Padua. Yet, if Capra had not later roused Galileo's ferocious indignation, we would think that he was merely scoring a point or two and otherwise more or less observing the courtesies of academic debate. It was not that he was one of those who was trying to preserve the heavens from change. It was thinkers more conservative than Capra who would feel threatened by the star. This new monster in the heavens would madden the Peripatetics, who had hitherto believed so many lies about the new star of 1572, which showed no motion or parallax. That is how Ilario Altobelli put it in a letter to Galileo of 3 November, knowing that Galileo would be looking forward to seeing the Peripatetic philosophers squirm in their struggles to show that the 'monster' was not really in the unchangeable heavens at all (10: 117).

Galileo catered for the widespread interest by giving three public lectures, probably in November 1604 (2: 525), of which only a few fragments and notes survive. At the beginning of the first lecture he coolly said that some had come because they were terrified and prompted by vain superstition to fear that the star was a portentous prodigy heralding all sorts of evil (2: 278); presumably the only reassurance they got was that Galileo did not share their worries. (Galileo never had any sympathy with alchemy or magic, and not much with astrology, though he did help friends with horoscopes or cast them for his family.) Others wanted to know whether the star was really in the heavens or just vapours near the Earth. Everyone was anxious to learn the motion, position, substance and cause of the apparition. Galileo's principal aim was to get the audience to understand that, since the star showed no parallax, it was certainly much further from Earth than is the Moon.

In the case we looked at earlier (annual parallax) the argument was about the seeming change in stellar positions one would expect to detect as a reflection of the Earth's orbit. It had been accepted from antiquity that a mere change of our position on the Earth itself could not lead to any such perspectival change, since we might as well be seeing the stars from the centre of the Earth. But it was well understood that heavenly objects that are nearer the Earth are a different matter.

When we see the Moon on the horizon there is a difference of nearly a degree between the position we see it in and its position as measured from the Earth's centre. Galileo, along with other astronomers, explained that the new star could not be as near the Earth as the Moon is because no such perspectival effect could be detected. In fact, it behaved like the old stars, which are too far away to show any such effect.[11] In a fragment of a letter of January 1605 to Altobelli, Galileo was rather apologetic about this: it was, after all, elementary stuff and hardly worth publishing. But he was mulling over some ideas of his own on the nature of the star, ideas which involved very great consequences, so he thought he would incorporate the lectures and these new ideas in a discourse for publication (10: 134–5).

This is tantalizing, since we cannot be sure what his ideas were. We know that he gave some thought to Copernicanism in connection with the new star. One of the fragmentary notes he made at this time is a quotation from Seneca to the effect that it would be worth knowing whether it is the Earth that moves or the heavens. There is also a suggestion that the annual motion of the Earth could be relevant to arguments over the new star of 1572 (2: 281–3). Incidental references to Copernicanism also occur in a *Dialogue* published in Padua in 1605, under the pseudonym of Cecco di Ronchitti (2: 309–34). This satirical dialogue, written in Paduan dialect, made fun of an inept work on the new star by an Aristotelian called Lorenzini, who was probably helped by Galileo's colleague, friend and adversary, Cremonini. In the *Dialogue* a practical man, a local surveyor, explains in homely terms what parallax is, showing that philosophers' attempts to arbitrate on the nature of the heavens were ignorant and ludicrous. This little work seems to be a joint production by Galileo and one of his pupils, Giacomo Spinelli (*c.*1580–1647), a Benedictine monk; it may even be, as Drake holds, that it is almost entirely Galileo's, with only incidental help from Spinelli. A few months later a second edition printed in Verona changed the favourable references to Copernicanism to slighting ones. All this makes plausible the conjecture that Galileo had been hoping that the new star would show some change of aspect that would reflect the annual motion of the Earth; when no such change was detectable, he had the favourable references to Copernicanism deleted. Even if this is granted, I doubt if much can be deduced from the episode about Galileo's appraisal of Copernicanism.[12] He can hardly have thought that failure to detect such a change of aspect in the new star *disproved* Copernicanism, though he may well have been disappointed that a possible proof of the new system had not emerged.

There is a further, unexpected, complication. Galileo was not at all sure that the new star was really a star at all. In company with all competent astronomers he put the new star well beyond the Moon: it was a heavenly object, no matter what Aristotelian philosophers might say. But he seems to have been much inclined towards what sounds like an Aristotelian explanation of the nature and origin of the star: his suggestion was that the phenomenon was caused by condensed vapours far out in space, vapours which may have had their origin on or near the Earth.[13] This may seem astonishing: how could a committed anti-Aristotelian make such heavy weather of leaving Aristotle behind? But it is astonishing only if we expect Galileo to have picked winners nearly all the time. In fact his tentative suggestions should not be called Aristotelian: he was, in his own way, introducing changeability into the heavens; he was making the heavens more like what we know on or near Earth. At the same time he stated quite roundly in the conclusion of his final lecture that, so far from knowing what the new star was, he did not even know the nature of the familiar stars, with the clear implication that no one else did either (2: 281). So it was quite reasonable to try any plausible way of accounting for the new star, though nothing could be put forward except very tentatively. This tentativeness, verging at times on scepticism, was to be seen again when the heavens once more provided matter for debate in the comets of 1618. On that occasion too Galileo took a position which has seemed to many perversely close to an already exploded Aristotelianism, with the extra twist that he also explicitly queried the relevance of arguments from parallax.

The Geometrical Compass

Galileo's first acknowledged publication was a little manual, of which sixty copies were printed in his own house in June and July of 1606. The manual contained instructions for using the geometrical and military compass which he had first produced in 1597 or 1598. In July 1599 he added to his household Marcantonio Mazzoleni, who made the instruments to Galileo's specifications (19: 131). The specifications became increasingly elaborate as Galileo hit on new ways to incorporate markings from which the solutions to a great variety of quite difficult problems in measurement could be read off. Since the compass was meant to ease the work of practical men such as military surveyors and gunners, some instruction in its use was needed. For years Galileo imparted this instruction in his private teaching. Many of his pupils were foreign gentlemen, so dozens of his instruments, with manuscript instructions, came to be scattered around Europe. An obvious danger of this was that they could be copied. Galileo gave as his reason for printing the booklet a suspicion that someone else was about to appropriate the instrument, a suspicion which turned out to be well-founded.

Galileo knew that similar instruments had been designed and produced quite independently of his. All he claimed in his letter to the reader was that most of the features of his version, and the most important ones at that, had never been hit on by anyone else. He insisted that oral instruction was necessary to make the manifold applications of the compass intelligible. The manual was for the initiate; it would never have been printed save to establish his intellectual ownership. The final point in his introductory letter can be seen as the merest common sense or also as a hint that he was prepared to look for an audience different from that which most professors catered for: since most of the mathematical applications of the instrument concerned soldiers, he had written in Italian. Whether one counts this as a published work or as a piece of private printing, it seems to have helped Galileo in August 1606 to clinch a renewal of his contract with the University with an increased salary of 520 florins (19: 114).

The plagiarism of this work by Baldassare Capra in his garbled Latin reworking of Galileo's ideas is the sort of idiocy that can be recorded but not explained. He did not even take the elementary precaution of publishing outside the Venetian Republic: in fact, he chose to publish in Padua itself and he added to the provocation by insinuating that the one guilty of plagiarism was Galileo. Clearly Galileo had to react to this wanton attack on his good name. Fortunately for him he could establish his essential claims with solid documentation and testimonials. He appealed to the law and turned the trial of Capra into a pitiless exposure of the man's ineptitude. There was no doubt of the justice of his case, so he obtained what he wanted, the total humiliation of Capra and the destruction of the offending publication. Immediately afterwards he published his version of the whole affair, ostensibly because a few copies of Capra's work had escaped abroad. The rhetoric is ferocious: even from his own account we can see that Galileo's strategy of annihilation seemed excessive to the judges, who had to restrain him from further demolition when his case was already well won. In this torrent of righteous indignation Galileo got away with a quite unnecessary and incredible claim: he would concede Capra's whole case and more if anyone could show that his instrument borrowed even the slightest detail from anyone else.[14]

There was no excuse for Capra's irresponsibility. Galileo had a perfect right, indeed an obligation to himself and his University, to expose his witless accuser. One does not have to share prickly contemporary Florentine notions of personal honour and reputation to sympathize with him. This was not like the later self-imposed exile of his friend Salviati over a petty matter of who should give way to whom in the street. But Galileo at his most ferocious was certainly alarming, even to some who wanted to see him achieve unqualified victory.

Motion Again

It is worth recalling from chapter 1 that Galileo's first attempt to become a research professor for the Grand Duke of Tuscany was made in February 1609, before he knew anything about telescopes. At that stage his case rested on his non-astronomical work, principally on mechanics and motion. Since none of that work was published he had to have recourse to self-advertisement. But he did offer more than the indefinite and high-sounding promises which we noticed in chapter 1. Another letter written at the same period (to Antonio de' Medici) is more forthcoming about what he had to offer. It is interesting to see how Galileo presents his novel discoveries in a way which would appeal to the Grand Duke. In particular, he has just completed all the demonstrations which concern the force and resistance of pieces of wood of different dimensions. This summary of what was to be the first of his two new sciences in the *Discourses* is naturally puffed as not only entirely novel but also essential in the construction of machines and of every sort of building. At present he is engaged in clearing up some remaining questions about the motion of projectiles – a study very relevant to the use of artillery, he points out. Then, lest his correspondent (or the Court) miss the urgency of closing with the proposal of such an original and practical mind, Galileo confides his very latest remarkable discovery, which he illustrates with a rough sketch. The sketch shows a series of paths of cannon-balls all rising to the same height: such paths, whether long or short, all take the same time. In a short letter Galileo can do no more than give a sample or two, but he looks forward to a conversation when he can explain three or four discoveries which are more wonderful perhaps than the greatest curiosities hitherto searched out by men (10: 228–30).

This is an excellent instance of Galileo's flair as a copywriter for advertisements describing his own discoveries. The temptation is to play down what he writes as the product of his habitual exaggeration and excessively good conceit of himself. But his letter was realistic in two respects. First, he knew that there was more work still to be done on the path of projectiles. Second, he was nevertheless confident that he had arrived at a firm basis for a new science of motion. Even the little we have seen in this chapter of his study of moving bodies is enough to suggest that his confidence was justified. The letter, no doubt, was written to promote his own career, but the reason he wished to obtain ideal working conditions was to further his work of original research. Since that research was to result in what can reasonably be counted as a new physics, anything less exalted than the language he chose to use in his letter might seem a little sheepish. As things turned out, his overtures to the Tuscan Court were ineffectual at this stage. As we saw in chapter 1, it was his use in 1610 of the newly invented telescope which gave him sufficient purchase to lever his way into the post he coveted. The fascination of astronomical novelties was great enough to distract him for the next twenty years from any sustained prosecution of the new understanding of motion which he had arrived at in those fruitful years from 1602 to 1609. But in all his astronomical work with the telescope and in all his defence of Copernicanism he had one immense advantage over many of his able peers: he not only understood that Aristotelian ideas of how bodies move presented no serious obstacle to a physics based on a moving Earth; he also had more than the rudiments of a very different mathematically based physics which Aristotelians could reject but not rival.

5

Discoveries and Controversies

In chapter 1 we saw Galileo setting out the reasons why he wished to leave Padua and become a research professor in Florence: he described an impressive list of works to be readied for publication, but he also assured the Grand Duke that he would make new discoveries. He made this promise good very quickly after his appointment on 10 July 1610. On 25 July, as he informed Belisario Vinta on the 30th, he noticed that Saturn is composed of three stars. Vinta and the grand-ducal family were to keep this secret until he had published it. The letter was to establish that he was the first to make this discovery and, of course, to impress his new employer (10: 410). Saturn continued to puzzle Galileo and everyone else. The solution came only in 1657, when Huygens explained that, though the planet seemed to have two close companions, which sometimes looked like ears or handles and sometimes disappeared, these appearances could be accounted for if it was surrounded by a ring. In Galileo's lifetime it was a puzzling oddity, but at least it was one which Galileo had spotted before anyone else, even if it did nothing to advance the cause of Copernicanism as at first he hoped it might. He further protected his priority by forwarding to Kepler an anagram of a phrase describing the discovery. In August he received a gratifying report that it was driving Kepler mad; it was November before Galileo wrote to solve the anagram and so make public this strange Saturnine phenomenon (10: 420, 474).

In his letter to Kepler in August Galileo was properly grateful for the singularly generous support Kepler had given him. What a contrast with the hesitations and obscurantism of so many at Pisa, Venice and Padua. What would Kepler say of the Paduan philosophers – Cremonini was one of them, though Galileo did not name him – who did not even want to see a telescope, let alone the Moon or the planets? Galileo dismissed them with an image which he later developed into the core of a manifesto: 'This sort of person thinks that philosophy is a book like the *Aeneid* or the *Odyssey*; truths are to be sought not in the world or in nature but (to use their words) by comparing texts' (10: 423). Kepler would laugh to see the primary professor of philosophy at Pisa using logical arguments as incantations to eliminate the new planets from the heavens.

Before long Galileo was to embarrass and browbeat the Aristotelians with a discovery much more significant than the strange shape of Saturn: his telescope revealed that Venus shows phases like the Moon's. That discovery was relevant to the comparative merits of Ptolemaic and Copernican astronomy. The very arrangement of the planets in Copernicus's system meant that Venus (and Mercury, if it were not so hard to see) ought to show such phases, whereas the traditional Ptolemaic scheme precluded them. In his *Dialogue* Galileo was to give Copernicus high praise for maintaining faith in his system, even when the required phases of Venus could not be observed. To praise such a triumph of reason over sense raises questions which must be postponed for a while. But if such phases could be observed it would be a very telling point in favour of Copernicanism and against Ptolemy. It would not prove that Copernicanism was the true system, but it would show that Copernicus was right in making Venus and Mercury orbit the Sun. Tycho, of course, had followed Copernicus to the extent of having all the planets in orbit round the Sun, while keeping the Earth, with its attendant Moon, fixed at the centre of the Sun's orbit. That arrangement was beginning to look attractive to many people precisely because it was a compromise, which combined most of the advantages of Copernicanism without the difficulties consequent on having the Earth move. In any case, a Ptolemaic astronomer who did not want to go as far as Tycho had could allow that Venus and Mercury orbit the Sun, provided the Earth was kept fixed at the centre of the universe. Borrowing such a major modification of Ptolemaic astronomy from Copernicus's system would seem to traditionalists like a sensible tactical withdrawal, made easier by the fact that it had already been proposed in antiquity. To Galileo, Tycho's system was no system at all and could never be turned into one, whereas Copernicanism was a system and one which held out hope of a celestial mechanics. To Galileo it was intuitively obvious that somehow the Sun must cause

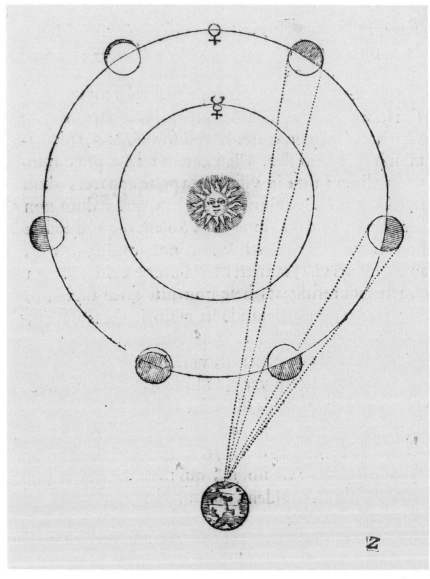

14   The phases of Venus depicted in Pierre Gassendi's *Institutio astronomica* (Paris, 1647).

planetary motions, so it would be physically absurd to insist on a special status for a fixed, central Earth. Hence the discovery of the phases of Venus was a wonderful vindication of *Copernicanism*; the fact that it fitted Tycho's arrangement equally well was to be brushed aside as irrelevant. The discovery also established that planets are opaque objects which shine only by reflecting sunlight, something which surprised even Kepler, as he confessed to Galileo (11: 78). But, to use a military metaphor, which is not inappropriate since before long Galileo was engaged in a campaign for Copernicanism, he did not assume that any war over the world-systems would necessarily go his way.

The fact that he had made another new discovery was reported on 11 December 1610 to Giuliano de' Medici in Prague, with another anagram to tantalize Kepler and an assertion that it brought with it the settling of very great controversies in astronomy: in particular it contained a very strong argument for Copernicanism (10: 483). It should be said that this was a discovery which was, as it were, waiting to be made. The Jesuits of the Roman College were observing Venus from November and in the course of December they recognized its phases independently of Galileo (11: 34). On 5 December Castelli had written to ask quite explicitly whether Galileo had seen such phases: a convinced Copernican himself, he hoped the answer would be yes and that it would win round doubters and opponents (10: 481–2). The question naturally arises: did Galileo merely follow up Castelli's suggestion? In other words, had the connection between Copernicanism and the phases of Venus simply escaped him until Castelli's query made it blindingly obvious that a major discovery was there for the taking? For this to be possible, as Drake points out, one would have to assume a very swift postal service between Brescia and Florence (probably via Milan in this case). Next one would have to have Galileo immediately appropriating the idea in his letter to Prague, admittedly only in the fairly safe form of an anagram, but an ingenious one which can hardly have been composed on the hoof. On these assumptions Galileo would have observed Venus himself before announcing the discovery at the end of the month, so he could still count himself as the discoverer of the phases, but he would be guilty of dishonesty in not giving credit to Castelli for prompting him to look for them.[1]

Yet it seems more likely that Galileo, like the Jesuits in Rome, had been observing Venus before Castelli's letter arrived. Galileo's account first to Clavius and Castelli in letters of 30 December, and later to others, stated that he has been observing Venus for three months. It is true that when he solved the Saturn anagram for Giuliano de' Medici in November he said that he had observed no novelty around (or concerning) the other planets, which was true enough if it referred to satellites. He could still have been observing Venus and biding his time until he was sure enough to announce that he had discovered its phases.[2] His claim was accepted peaceably and Castelli did not demur. So this wonderful year was ending very happily for Galileo: not only was he sending to the influential Jesuit astronomer news of a great discovery, he was replying to Clavius's of 17 December which reported that at last the Roman Jesuits had managed to see the satellites of Jupiter, though Saturn seemed to them oblong rather than made up of three stars; let Galileo continue to observe, concluded Clavius, perhaps he would make new discoveries about other planets. (Perhaps Clavius had in mind the phases of Venus, but was not himself ready to declare himself on this topic to Galileo.)[3]

Soon Galileo did make a new discovery, though for him the next one was not a discovery about a planet, but about what he was now quite sure was the centre of planetary motions, the Sun. That discovery would bring him into difficulties, partly of his own making. But already there had been enough opposition to his *Sidereal Message* to make him realize that to establish Copernicanism would not be easy. On the day he wrote to Clavius he replied warmly to Castelli to report that Venus does indeed show phases and that great consequences can be drawn from that and his other discoveries. But Castelli's suggestion that this discovery would convince the obstinate had almost made him laugh. Did Castelli not know that the demonstrations produced earlier were sufficient to convince those who can reason and want to know the truth, but that to convince those who are obstinate and care only for the empty applause of the stupid mob the witness of the stars come down to Earth to speak for themselves would not suffice? Let them endeavour to gain knowledge for themselves, but give up hope of making headway in popular opinion or of convincing bookish philosophers (10: 502–5). The evident bitterness in this letter shows a side of Galileo's character that from now on appears repeatedly in his private writings and before long in his published or semi-published work. When engaged in public controversy he can seem arrogant and unstoppable, at least by argument. This letter suggests that he was also very vulnerable and actually hurt by reluctance or even refusal to accept his discoveries. It also seems to say that, even before the discovery of the phases of Venus, adequate proof of Copernicanism had been available: some of Galileo's troubles were to derive from this conviction. To the end of his life he expected people to be convinced by a very simplified version of what Copernicus had put forward. Still, he had won a great victory for the new philosophy and reported it as such to the Tuscan ambassador in Prague and through him to Kepler on New Year's Day 1611 (11: 11–12).

Roman Triumph

On Holy Thursday, 29 March 1611, Galileo arrived in Rome to stay as the honoured guest of the Tuscan ambassador. He would see St Peter's not only with Michelangelo's dome at last built by Carlo Fontana, but lengthened by its vast nave, with an imposing façade by Carlo Maderna nearing completion. The purpose of his visit was to justify the complete truth of all he had discovered: this purpose was endorsed by the Grand Duke in his letters of introduction to cardinals, who in any case were eager to see him. The day after his arrival Galileo had a long conversation with Clavius, Grienberger and Maelcote. It was heartening to find these Jesuits laughing at a recent attack on Galileo by Francesco Sizzi, though embarrassing that the inept author was a Florentine (11: 79–80). Galileo thoroughly enjoyed being lionized by cardinals, prelates and princes.

One banquet at which he was the guest of honour had important consequences for his work. It was given on 14 April by the Academy of the Lynxes (the Accademia dei Lincei), a society which had been founded by Federico Cesi, Marquis of Monticelli, in 1603. Academies were a notable feature of literary and artistic life in Italian cities. What was unusual about Cesi's was that its members were not to confine themselves to elegant literature, but were to give pride of place to the study of nature and mathematics. It is true that Galileo was being honoured by a tiny society which had as yet done little to realize Cesi's hopes, largely because of the obstructiveness of his father. But Cesi became independent in 1610, when the administration of the family's property was made over to him. (From January 1613 he was styled Prince Cesi, the title being a papal honour.) The banquet marked the real beginning of the society: Galileo and other able men among the guests were soon admitted as members and the dormant Academy turned into what may be counted as the first successful scientific society. Its fame largely depended on Galileo's achievements, but to Cesi goes the credit of seeing the need for cooperative endeavour and the support of research and publication. He financed it all; he did all the necessary organizing. If the Academy gained glory through Galileo's writings, in return it gave him much-needed support, stimulus and encouragement. He had left the congenial friends of Padua and Venice; there were a few in Florence who appreciated his studies, but there were many more who could not even understand what he was doing. Galileo was proud to be a Lyncean and took full part in the correspondence which was the Academy's main form of cooperation. It is fitting that the instrument which gave him fame was first called 'telescope' by Giovanni Demisiani, one of his fellow-guests at that banquet in Rome, or by Prince Cesi himself (11: 420).

Naturally, the Tuscan ambassador presented Galileo to Paul V and made the occasion more memorable for Galileo by explaining that the Pope had treated him with unusual signs of favour (11: 89). A couple of days later Cardinal Bellarmine inquired of his Jesuit colleagues at the Roman College what they thought of Galileo's observations (11: 87–8). They replied on 24 April. The telescope had certainly revealed stars hitherto invisible, though it was doubtful whether the Milky Way is composed entirely of tiny stars. Saturn looked egg-shaped and oblong, though the two little stars on either side had not been seen detached from it. Venus does show phases. Clavius thought it more probable that the Moon's surface is really smooth, while others thought it is uneven, but they were not certain enough to say the matter was beyond doubt. Jupiter does have satellites (11: 92–3).

As I mentioned in chapter 2, it is appropriate to call Cardinal Robert Bellarmine by the Anglicized version of his name since he was something of a bogeyman in England at that time, partly as the opponent in controversy of James I, partly as the leading theologian of the Counter-Reformation, and partly, perhaps, as the archetype of the Jesuit guile and cunning attributed wholesale to the Society by opponents, both Protestant and Catholic. Modern scholarship has treated Bellarmine pretty fairly on the whole. His active role in the Church's first moves against Copernicanism has usually been presented as an honourable, if mistaken, attempt to find a compromise acceptable to all legitimate interests. In the next chapter we shall see his preoccupations and his proposed solution to the challenge presented by Copernicanism, but he was such an influential figure in the Church that his interest in Galileo's discoveries merits attention.

One naturally wonders whether Bellarmine was checking up unofficially to see whether the discoveries everyone was talking about raised difficulties for theology. It could be that he was indulging his known amateur interest in astronomy, although a disturbing reply would doubtless have alerted him to the need for further investigation. Now the reply, as we have seen, was cautious but in no way alarming: it was certainly a more than adequate endorsement of Galileo's discoveries. Yet Bellarmine was present at a meeting of the Inquisition on 17 May which minuted: 'see whether Galileo is named in Cremonini's case' (19: 275). It is undeniable that Galileo kept what Bellarmine would have considered bad company: Cremonini, no doubt, was just an awkward Aristotelian of dubious orthodoxy; Sarpi, who had only recently been pitted against Bellarmine as Venice's theological champion in the struggle between the Republic and the papacy, was quite beyond the pale. One of the attractive features of Galileo was the range of his friendships, but it would have seemed suspect to Bellarmine. So one could use the Inquisition's minute as a starting-point for a variety of interesting conjectures. Unfortunately there is no knowing just what the Inquisitors had in mind. Perhaps they felt that a routine check on someone who was spreading such novel ideas could do no harm and might detect dangers hitherto latent. Bellarmine himself, according to what the Tuscan ambassador to Rome wrote four years later, became edgy as Galileo was fêted all over Rome and was considering doing something about it (12: 207).

The Jesuit mathematicians evinced no such qualms. Their caution in accepting his discoveries had exasperated Galileo, but it was what was required by their *Ratio Studiorum* and, in any case, needed no excuse. Once they were convinced he was right they showed their appreciation handsomely and publicly in a lecture, given at the Roman College in May by Odo van Maelcote, in the presence of Galileo himself. The audience of the learned and notable included counts, dukes and three cardinals. Maelcote even moved much nearer to accepting Galileo's description of the Moon's surface; we may presume that it was respect for the aged Clavius that made him say politely that he was merely reporting appearances and that some might wish to appeal to the varying density of the Moon's sphere.[4] The message was clear: the Jesuits were endorsing Galileo's discoveries. The occasion was intended as a triumph for Galileo, a triumph that Mark Welser, a banker from Augsburg who was beginning to correspond with Galileo and was just the sort of person who needed some endorsement from authoritative persons to settle his opinions, optimistically thought would snuff out any remaining spark of envy (11: 127). In the midst of all this excitement, perhaps the most remarkable aspect of Galileo's hyperactive three months in Rome is that he did not neglect astronomical observation. In fact he did something Kepler had thought impossible: he worked out, in a first approximation, the periods of Jupiter's satellites.

Floating Bodies

Galileo returned triumphantly from Rome in June 1611. The Grand Duke had every reason to be pleased with his mathematician. But Galileo had also insisted on the title of philosopher and was eager to

15   The Jesuits' Roman College (Gregorian University) from Giuseppe Vasi's *Delle magnificenze di Roma antica e moderna*, book 9 (Rome, 1769).

keep his promise of debating with philosophers. He knew that he had more than the beginnings of the new science of motion, which would be necessary to defend a moving Earth. More generally he knew that any campaign for Copernicanism would also have to be a campaign against Aristotelianism. Late July saw him engaged in informal discussion with Aristotelian philosophers, probably in the palace of Filippo Salviati, the other friend whom he immortalized in his *Dialogue* and *Discourses*. This discussion was to develop into a significant confrontation between Galileo and local Aristotelians; the ensuing published controversy had marked undercurrents of rivalry for patronage at the Tuscan court. The topic of the original conversation was condensation and rarefaction. An Aristotelian, probably Vincenzio di Grazia, professor of philosophy at Pisa, maintained that cold was caused by condensation, as is clear from the fact that ice is condensed water. To his surprise Galileo said ice seemed to him to be rarefied water, witness the way it floats on water. It only floats because of its shape, was the reply. Shape has nothing to do with it, retorted Galileo: all that matters is whether objects placed on water are denser than the water. Thus the scene was set for further arguments.

Three days later Lodovico delle Colombe, an amateur philosopher who had disagreed with Galileo over the new star and also circulated a manuscript against the motion of the Earth, entered the debate. He undertook to refute Galileo's claim that shape had nothing to do with whether bodies float or sink. There followed a flurry of rival forecasts about what would happen when bodies of varying shapes were placed in water, challenges to debate the matter and disputes about who had altered the original terms of the debate or failed to turn up at the appointed time. At the end of September there was even an informal debate in the presence of high personages, including Cardinal Maffeo Barberini, who sided with Galileo, and Cardinal Gonzaga, who favoured the Aristotelians. The Grand Duke thought things were getting out of hand and advised Galileo to deal with the topic in writing. So Galileo wrote down his views for Cosimo and then treated the whole topic of floating bodies in a *Discourse* which appeared in May 1612 and went into a second edition by the end of the year. As the Grand Duke might have foreseen, it provoked four lengthy replies. This published controversy made plain to all that Galileo was attacking Aristotelianism and that opposed to him was an informal group of Aristotle's Pisan and Florentine defenders; this group was nicknamed 'the pigeon league' after its most prominent member, Lodovico delle Colombe, since *colombo* means 'pigeon' and a secondary meaning of 'pigeon' was 'simpleton' (or 'bird-brained').

The fact that Galileo chose to write in Italian was itself an indication that he was dissociating himself from the Aristotelian philosophers and appealing to an audience wider than theirs. He explained to a friend that he wanted everyone to be able to read what he wrote. He had noticed that many of the young men who studied medicine or philosophy at universities were quite unsuited to such studies, while others with no such opportunities but plenty of natural ability told themselves such things were not for them. Galileo wanted these latter to realize that nature had given to them, as well as to philosophers, eyes to see her works and brains to understand them. To write in Italian was also, of course, to expose the pretensions of mere commentators on Aristotle, who clung to seeing things his way rather than looking and thinking for themselves (11: 327).

Perhaps the most significant aspect of the controversy is that it was seen to be about more than floating bodies: apart from competition for the court's favour, the question was whether physics should remain Aristotelian or should take Archimedes, in the person of Galileo, as its guide to method. (In fact, Galileo was going beyond Archimedes with the help of the pseudo-Aristotelian *Mechanics*.) An anonymous contributor to the debate, maliciously but perceptively accused Galileo of empire-building, of wanting to add a terrestrial empire to the one he had already claimed in the heavens (4: 156). It was not a very satisfactory controversy. Galileo had no difficulty in developing what he had written about specific weight in his manuscripts on motion and refuting, quite courteously it must be said, what his opponents had argued orally. Where he made advances, as Shea points out, he was arguing not against his opponents but against the over-simplified hydrostatics he had himself been satisfied with up till this time.[5]

The trickiest problem was what to say about little slivers of ebony carefully placed on water by his opponents: why did they not sink, as Galileo's theory demanded? The Aristotelians thought this showed the resistance of water to division, but Galileo dismissed this. Neither side suspected what much later work would reveal about surface tension, so Galileo acquitted himself fairly adroitly by pointing out that the slivers only floated if thoroughly dry, in which case they sat in a little depression in the water. In working out the effective specific weight of a sliver one had to add to its volume that of the air in the depression, so that an equivalent volume of water would weigh the same as the sliver.

He also applied himself to studying the published replies to his book, but was glad when Castelli took over much of the tedious chore of drafting an answer (which included the lapidary assertion that it is impossible to understand Copernicus without agreeing with him (4:

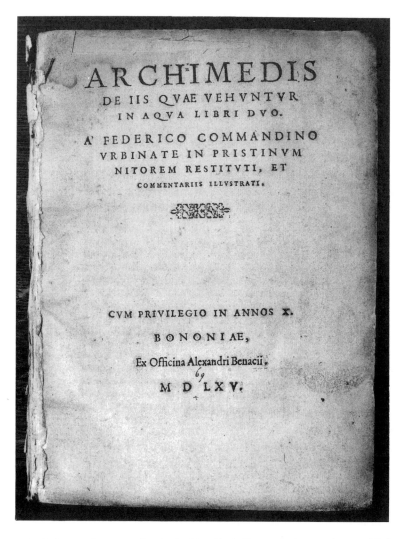

16   Federico Commandino's *Archimedis de iis quae vehuntur in aqua libri duo* (Bologna, 1565) was part of the sixteenth-century recovery of Archimedes, a tradition which Galileo developed further with his own work on floating bodies.

589)). It came out in 1615, with no name on the title-page but Castelli's at the end of the dedication, which is dated 2 May (4: 451–691). By that time Galileo had more important things to worry about. In any case, it is hard to disagree with his conclusion that he had convinced those who could understand what he was saying and it was not worth spending further time on those who could not. No one should take that as proof that he was arrogant without trying to read the works of his opponents in this controversy. The crucial blockage, as Galileo realized, was over mathematics: if he could teach them mathematics, he could convince them; otherwise there was no hope (4: 443). Whether this makes his views and theirs strictly incommensurable, to use a concept that has appealed to many in recent decades, is not clear. The debate in fact established little common ground of any significance, but if Galileo's diagnosis is right, he was not requiring a sudden, rationally unjustifiable, switch to his way of explaining things. On the contrary, he was convinced that virtually anyone who took the trouble could acquire the necessary intellectual tools to assess what he was saying. This problem of seemingly incommensurable views will come up again.

Sunspots

Disputes about priority of discovery are a significant feature of Galileo's career – and of the careers of some other scientists, notably Newton. In most of Galileo's disputes the evidence, though patchy, is extensive enough to keep controversy alive even today. It is tempting to relegate this topic to the occasional footnote rather than give it space that could be allotted to Galileo's scientific thought, but the way much of his thought developed was through controversy. When such controversy turned into disputes about who first made a discovery Galileo was not always an innocent victim, as he had been in the case of the geometrical compass. In his later works he wished to be accepted as the sole discoverer of all the great novelties revealed by the telescope. This was not a crazy claim, but it was inaccurate and not essential to the unique position he undoubtedly holds in early work with the telescope. It left little room for the independent contributions of others, whose achievements were slighted or treated as little better than a following of his lead or even plagiarism. The discovery of sunspots cannot be dissociated from Galileo's exchanges with the Jesuit Christoph Scheiner (1573–1650), since in many accounts of Galileo's later sufferings Scheiner haunts the background as a sinister figure.

Theirs was a European squabble: neither knew that the Chinese had been familiar with sunspots for centuries. Quite independently of both of them other European astronomers noticed spots in the sun: Johannes Fabricius actually published a pamphlet on the subject about June 1611;

his observations seem to have been made in March. Thomas Harriot in England recorded his first observation on 18 December 1610.[6] There was no reason why several people should not make such a discovery, once the telescope came into use. Welser, the intermediary for Galileo and Scheiner, realized this. That Galileo made his discovery independently of Scheiner is not open to serious doubt. By April 1611 he was showing sunspots to notable people in Rome. It is not easy to say when he first noticed them: the claim in the *Dialogue* that he showed them to people before he left Padua in 1610 was made in good faith, but relied entirely on the recollection of his partisan supporter, Fulgenzio Micanzio, the biographer of Sarpi and an enemy of the Jesuits (14: 299). Micanzio's account is circumstantial enough but it has to be viewed with some suspicion, since no claim to a Paduan sighting was included in Galileo's *Letters on Sunspots*. Nor can we derive anything precise from Galileo's loose and inconsistent claims in letters about how long ago he first saw them. But that he had seen them by early 1611 seems likely enough, though mere noticing is not the same as careful observation.

Scheiner's first serious *observations* were made only on 21 October 1611, but in reporting them in November he said he had first noticed sunspots about seven or eight months previously. In his definitive work on the topic, published in 1630 (though this part was written in 1627), he says more precisely that he first saw them, in company with a colleague, in March 1611 and insists that he had had no prior notice of them from anyone. This is perfectly credible. The only thing that can cast doubt on it is a report that his fellow-Jesuit, Paul Guldin (1577–1643), claimed very emphatically that he had been the first to tell Scheiner that spots are visible in the Sun and that Galileo was their first discoverer; this reported remark was passed on to Galileo in 1635 (16: 189; 17: 296–7). Guldin was as certain as could be of this, even after nearly twenty-four years. His Copernican sympathies and admiration for Galileo were well known. His flat, and quite irrelevant, disavowal of Copernicanism in the preface of a book he published in 1634, can reasonably be attributed either to understandable caution after Galileo's condemnation or to compliance with an instruction from his superiors, especially since he sent a copy to Galileo (16: 398; 17: 193).[7] Guldin, if he has been reported accurately, must have known that he was impugning Scheiner's published account. On this evidence it is not easy to choose between his version and Scheiner's, though if Scheiner did first notice the spots in March 1611 then Guldin's case collapses, since Guldin cannot have learned about them from Galileo before April.

Even Scheiner's report of how he came to make his observations in October has been queried: in his *Rosa Ursina* he said that, because of the favourable conditions on 21 October 1611, he decided to make observations of the sunspots he had noticed months before. But it was known from an independent account that almost at the same time in another room of the same house the Jesuit theologian, Adam Tanner, started to observe them, moved to do so by a report of their existence from elsewhere. Tanner's innocent excitement still comes over to the reader in his published accounts and is a useful reminder that even very conservative theologians could welcome telescopic discoveries.[8] (Tanner's name is to be found on maps of the Moon. This honour does not commemorate any significant contribution he made to astronomy; it is just a mark of esteem from his fellow-Jesuit, the astronomer and cartographer Riccioli.) Scheiner referred to Tanner's report but insisted that he had heard no such rumour before his earlier sightings of sunspots in March. It is hard to believe that 'some rumour or other' could set the theologian to using a telescope without the astronomer in the house being aware of it. But the rumour, whatever it was, need not have had anything to do with Galileo; in any case it would still leave Scheiner's claim about his March observations exactly where it was, so we are left with his word against Guldin's. I find Scheiner's claim believable because his first description of his first sighting says that he and his companion noticed sunspots only incidentally while they were comparing the apparent sizes of Sun and Moon and that they made little of them at the time, postponing further investigation till later (5: 25). It is very likely that at that stage Galileo himself had not done much more; there is certainly no evidence that he had, a fact which Scheiner quite reasonably exploited to the full in his *Rosa Ursina*.

The dispute concerned more than priority of discovery. If there really were spots on the Sun, then it would be less plausible than ever to maintain the essential Aristotelian contrast between heaven and Earth. Nevertheless, Welser encouraged Scheiner to publish; since it was a topic that would certainly interest Galileo, Welser was willing to act as intermediary. Busaeus (Jean Buys), the Jesuit Provincial in charge of the Society in that part of Germany, was pleased by Scheiner's discovery but thought the topic so novel and controversial that it would be wise of Scheiner to publish under a pen-name; in this he was merely exercising the caution the General required of all Jesuits. Hence the title: *Three Letters on Sunspots (Tres epistolae de maculis solaribus)* has the rather awkward author 'Apelles concealed behind the picture', alluding to the great Greek painter who hid behind his paintings to pick up uninhibited criticism. Apelles welcomed expert criticism from a cobbler on the subject of footwear, but had to come out and tell the man to stick to his last when he got too cocky. Scheiner was just as genuine in inviting criticism but also got more than he had bargained for. He was to become the acknowledged authority on sunspots but that does not alter the fact that when Galileo got round to replying he outshone Scheiner completely.

Scheiner's short Latin work was published on 5 January 1612 and sent to Galileo the next day by Welser. In the course of the year Galileo sent Welser two very full Italian letters in reply and then a third in which he took account of Scheiner's *Accuratior Disquisitio* ('More Accurate Disquisition'), sent to him at the end of September. Galileo's letters were published in March 1613 by the Lyncean Academy, with part of the edition incorporating Scheiner's two works, which were hard to find in Italy (5: 10–12).

After all the lingering controversy about the original dispute over priority it is a relief to record that there is not much argument about whose views on sunspots were better at that stage: Galileo wrote courteously, if a trifle condescendingly at times, but his work had a masterly sureness of touch and a genuine unwillingness to claim greater knowledge of sunspots than was available. Scheiner too wrote tentatively but he was hampered by a desire to preserve Aristotelian teaching unless there was conclusive proof that it should be rejected. Rather than have spots on the Sun, which would make a heavenly body imperfect and changeable, he preferred them to be real bodies, like little planets, between us and the Sun. It is interesting to see how Scheiner and other Aristotelians, once they had accepted the satellites of Jupiter, were willing to see similar bodies revealed by the telescope, because no change in the heavens was involved. Galileo, who was glad to have another means of exploding the Aristotelian prejudice about the unchangeable and incorruptible heavens, took great care to show that if, as he had discovered, the Sun rotates monthly on its axis, then careful consideration of perspective shows that the spots must be on or very near the Sun: the appearances of impermanent spots would be just as we see them, with some longer-lasting spots brought back into view by the Sun's rotation. They are not like little planets between us and the Sun; if they resemble anything familiar it would be clouds. Though the spots are dark in contrast with the rest of the Sun, in themselves they are brighter than any part of the full Moon. On all these points he corrected Scheiner and was genuinely surprised that Scheiner used laborious arguments about the visibility of transits of Venus across the Sun in an attempt to settle whether Venus orbits the Sun rather than the Earth: had Scheiner not heard of Galileo's discovery of the phases of

Venus, which put the matter beyond doubt? Scheiner caught up with this in his *Accuratior Disquisitio* in a part dated January 1612, so the omission in the first publication was an oversight, though one that showed he could not see the significance of the phases as clearly as Galileo had. But he added a curious claim to have discovered a fifth satellite of Jupiter, a claim which Galileo demolished briskly, thereby damaging further Scheiner's attempt to treat sunspots as similar to the satellites.

Galileo's attribution of rotation to the Sun itself led to some incidental reflections on the problem which was never far from his mind: motion. Heavy bodies fall naturally if unimpeded but need an external mover to take them upwards. But a body on a spherical surface concentric with the Earth will be indifferent to rest or motion towards any point of the compass, so if at rest it will stay put and if in motion, say westwards, it will continue in that motion. Thus in a calm sea a ship would stay at rest till moved by something and once put in motion would sail for ever. This view of motion is nearer to Newton than to Aristotle.

In addition to defending his views on earthshine against Scheiner and giving a tutorial on Castelli's neat way of viewing the Sun safely by projecting its telescopic image onto paper, Galileo naturally took the opportunity to instruct the world on the free way of philosophizing. Apelles is given credit as a free not servile mind, well able to understand true teaching and accept the true and good philosophy, particularly concerning the constitution of the universe. It is unfortunate that he cannot quite detach himself from ingrained fantasies, as is shown by the fact that he seems to take seriously the spheres and epicycles of the Moon, Venus and Mercury. Such devices are used by calculating astronomers to make predictions, whereas philosophical astronomers also want to investigate the greatest of all problems, the true constitution of the universe, 'because there is such a constitution, and it is unique, true, real and cannot be otherwise'; because of its greatness and nobility this question should be placed by speculative minds before anything else that can be known (5: 102). This cannot have pleased Scheiner: he undoubtedly saw himself as open to new ideas, since the motto at the end of his letters said that what is known is the least part of what remains to be found, and in the *Accuratior Disquisitio* he expressly rejected the solidity of the heavenly spheres. He even quoted from Clavius's recent edition of *On the Sphere* a remark that Galileo was often to use, namely that the new discoveries meant that a new system of the universe was needed (5: 69). This may actually be a response to Galileo's remarks, since it was written after Galileo's first letter had reached Welser, but it does not contain anything that Scheiner could not already have thought out for himself. Neither Clavius nor Scheiner was advocating Copernicanism, nor would they have accepted Galileo's concluding remark that Copernicanism was experiencing such favourable winds that there was little ground to fear crosswinds (5: 238). Clavius was dead by the time Galileo's *Letters* were published, but Scheiner was a potential ally. Had Galileo's eagerness to promote Copernicanism and the right way of doing science alienated Scheiner?

In the body of the text there are things which would not convince Scheiner and things which would displease him, but it does not seem a particularly acrimonious exchange. Now the Lyncean Academy was Galileo's publisher: the members, especially Cesi, went to a good deal of trouble to satisfy the censor by excising the perfectly sound argument that scripture does not seem to favour the immutability of the heavens (5: 139). This, incidentally, was an argument which Bellarmine had accepted all his adult life and Scheiner's *Rosa Ursina* was to buttress with a massive accumulation of references (plus references to Cesi's and Bellarmine's opinions). Galileo's well-placed gifts of telescopes entitled him to seek counsel from cardinals. In July 1612 Cardinal Conti had advised him that, so far from the scriptures favouring Aristotle on the incorruptibility of the heavens, the common opinion of the Fathers was that the heavens are corruptible. But to use the new discoveries to prove corruptibility could be difficult and he would like to hear Galileo make out a fuller case. As for claiming that the Earth moves, that would involve reinterpreting scriptural passages as merely conforming to our common way of talking: such a reinterpretation could not be allowed without compelling necessity; only Diego de Zuñiga in his commentary on Job had adopted this interpretation. Even after further explanations from Galileo in August, Conti evidently preferred Scheiner's view of sunspots as starlets. His two letters give a fair indication of the state of influential Church opinion in the summer of 1612: they are an early articulation of difficulties which Galileo would soon have to face (11: 354–5, 376). It is interesting that the censor passed a favourable reference to Copernicanism, but would not allow Galileo to argue that scripture showed the heavens are corruptible, presumably because that would be to stray into theology.

It would seem, then, that Galileo was not being unduly controversial in what he published. But the Academy's librarian wrote a letter to the reader which was more combative than the text it introduced (5: 83–4). The published version is milder and more cautious than the extant manuscript, presumably because Galileo had the good sense to tone it down. But what was printed not only set out to establish how early Galileo had observed sunspots, but cried up his unique role in making and exploiting telescopic discoveries. The letters would be welcome to everyone, except perhaps some who either made claims about having discovered them or wished to maintain their own opinions about them or found these novelties too threatening to what they have taught as certain and so were unable to accept with open minds what came simply from Galileo's pure desire and concern for truth. It is possible that others saw sunspots without prior notice from Galileo, but he certainly showed them in Rome in April 1611 and months before that to friends in Florence, whereas Apelles's first observations belong to October of that year.

This may have been published against Galileo's better judgement, but it did have his consent and it set the pattern for future writings. One does not have to be a partisan of Scheiner's to see how it might have alienated him. He had after all published his observations well before Galileo and sent him his publications. Yet he does not seem to have reacted strongly at this stage. He may have read the preface as claiming only that Galileo's showing of sunspots in April was independent of, but after, his own noticing of them in March. Perhaps for this reason he tried to keep contact with Galileo.

That he did try to maintain contact was no thanks to Sagredo, who had forwarded Galileo's letters to Welser. In April 1614 Sagredo informed Welser that he would no longer exchange views with Apelles. The gist of his letter was that it was not worth the while of a Venetian gentleman to deal with a pretentious and captious writer, who could not even admit the date-line paradox (which Sagredo did not pick up from Galileo but spotted himself), but insisted that the whole world could celebrate a feast on the same day. Since Sarpi and other mathematicians (and possibly even Galileo) were just as reluctant to concede Sagredo's point, the contempt for Apelles must rest on the Jesuit's inability to observe the ordinary courtesies of debate (11: 554–5; 12: 45–6). This, of course, is only Sagredo's side of the question and he was acting for himself, not Galileo; but the breach must have been embarrassing to Welser. Sagredo informed Galileo that he had broken off correspondence and said he believed Apelles was the Jesuit François d'Aguilon; presumably he was misled by the fact that d'Aguilon had just published a work identifying Apelles as Scheiner, a fact which Cesi had told Galileo about at the beginning of March (12: 29, 51). Scheiner may not have known anything of this, as Welser died in the summer. Scheiner did write Galileo two polite letters in February and April 1615, enclosing in each a minor publication and asking for criticism (12: 137–8, 171–2). Galileo did not reply.

An important contrast between Galileo (and the Lynceans) and Scheiner (and the Jesuits) was already evident. The Lynceans, through Galileo, advocated a free way of philosophizing or, as we might say, of doing science. They were confident, as we shall see in the next chapter, that their free way need not lead to conflict with the revealed truths of religion. The Jesuits were just as convinced that no truth could contradict another, but they were as a matter of policy much more cautious about accepting claims that a new truth had been discovered. We have already seen that their *Ratio* prescribed such caution and proscribed novel opinions. What it did not do was give guidance on how a novel opinion could come to be accepted: if Jesuits had to wait for it to be accepted by universities, academies or learned authors, they would be confining themselves to the role of traditionalist critics, whereas hitherto they could fairly be called cautiously open to new ideas and even willing, as Scheiner was, to publish novelties themselves.

Now it so happened that the controversy over sunspots took place in a period when the Jesuit General, Claudio Aquaviva (or Acquaviva), was insisting on ever greater caution in the implementation of the *Ratio*, which he naturally saw as one of the great achievements of his long generalship. He addressed two letters to the Society, one of 24 May 1611, the other of 14 December 1613: the thrust of them was to insist that the provisions of the *Ratio* be carried out strictly. Aquaviva's unease was probably caused by a fear that irresponsible Jesuits would reopen recent theological wounds, in particular the dispute with the Dominicans over grace and free will. That dispute would have been endless, had not Paul V, at Bellarmine's suggestion, imposed a sort of truce on the disputing parties in 1607.[9] The last thing Aquaviva wanted way anything that might look like a breach of the cease-fire by a Jesuit. But his strictures applied to philosophy, as well as to theology. We need not think that he was opposing Galileo's discoveries reported in the *Sidereal Message*: the Roman College could hardly have staged their celebration of Galileo at the very time Acquaviva was writing his first letter if that had been its meaning. (It is, admittedly, just possible that Aquaviva was reacting to the Roman College's endorsement of Galileo, but I do not know of any evidence to support such a conjecture.) But even the first letter was sufficient to make Jesuits like Scheiner and Grienberger extra cautious. Scheiner, we have seen, had to use a pen-name. Grienberger acknowledged quite openly to Galileo himself that he was not a free agent. He sent Galileo a letter on 5 February 1613 about a mathematical problem. If Galileo chose to send it on to Kepler, that would please Grienberger. Then he concluded: 'Do not be surprised that I say nothing about your [writings]: I do not have the same freedom as you do' (11: 480).[10]

6

The Condemnation of Copernicanism

The *Discourse on Floating Bodies* and the *Letters on Sunspots* confirmed to the reading public what had been clear enough in *The Sidereal Message*: Galileo meant to discredit Aristotelian physics and cosmology wherever he could. He was better prepared than most people could have realized. He had started to expose the inadequacies of Aristotelian accounts of motion long before he made his telescopic discoveries, very likely before he ever accepted Copernicanism. It is true that from the middle of 1609 the telescope diverted him from the study of falling bodies and projectiles but, as a few friends knew, not before he had discovered the basis of a new science of motion. He was in an excellent position to start writing the work on the system of the world which he had promised in his *Sidereal Message*. There would never be a better opportunity to promote the free way of philosophizing, free from the authority of Aristotle (or anyone else), free to understand nature by means of sensory observations and conclusive proofs. There was indeed little hope of winning over to his side the local league of professional and amateur philosophers who had opposed him tenaciously and sourly over floating bodies. Able Jesuits, however, like Grienberger, Guldin and Scheiner were potential converts, if only they could be brought to see that their policy of following Aristotle as far as possible was no longer defensible. If they could be recruited, the influence of their Society would transform the teaching of philosophy in Catholic institutions worldwide. Nor was Galileo a solitary campaigner: he had the support of the Lyncean Academy and of its leader, Prince Cesi.

The Academy itself did not espouse Copernicanism but it did spon-
sor the *Letters on Sunspots* and it was committed to the approach to
science exemplified in them. In May 1613 Cesi was looking for 'soldiers
in our philosophical militia' and he asked Galileo's help in recruiting
men 'of free minds in natural philosophy'. A casual mention confirmed
the nature of the Academy's most distinctive pursuits, namely profound speculations in physics and mathematics (11: 507). Cesi was
often a calming influence on Galileo in polemical mood, but he also
consistently encouraged him to promote the free philosophy which the
Academy was founded to pursue.

Another support to Galileo was Castelli. He was no less zealous than
Galileo for Copernicanism and just as forthright in his attacks on
Aristotelianism. He was also a very able collaborator and a willing one,
as he showed by taking over the drudgery of replying to the tedious
cavils against the *Discourse on Floating Bodies*. In the autumn of 1613
Galileo heard the satisfying news that his recommendation had secured for Castelli the chair of mathematics at Pisa. It was no surprise to
Castelli to find that this was largely enemy territory. On his arrival
there the administrator warned him not to deal with the motion of the
Earth. Castelli replied that he would, of course, comply, the more so
because Galileo, who had never dealt with the topic in twenty-four
years of lecturing, had given him the same advice. If this left the
administrator slightly flustered he managed to reply that Castelli might
well have used occasional digressions to treat similar questions as
probable. The well-prepared Castelli smoothly promised to avoid that
too unless the administrator instructed him otherwise. All this he
reported to Galileo on 6 November (11: 589–90).

A week later he confided to Galileo 'as a father' that his inaugural
lecture had been a tremendous success. At the same time he was
startled to find how little was known in Pisa of Galileo's discoveries
and he had already had an encounter with an angry challenger whose
ignorance of Euclid was comic (11: 594). On 20 November he reported
that Galileo was envied not for his discoveries, which were unknown to
the malicious, but for his exceptionally high salary (11: 596). These
snippets help to describe the context of the first significant attempts to
use theology and the authority of the Church to oust Copernicanism
from philosophical discussion. In themselves, though, they were part
of the stuff of university life: Borro had been pushed out of his post
decades earlier by rivals and Buonamici's massive tome, as he explained, arose out of local controversy among staff and students on the
motion of the elements.[1]

For three and a half years Galileo had skilfully deployed his telescopic discoveries to discomfort Aristotelians. He had allowed himself the occasional public hint that Copernicanism was the true system of the universe, but he had avoided any public entanglement with theology. In the dispute over floating bodies Colombe had aired an interesting grievance, namely that Galileo had never replied to the tract *Against the Motion of the Earth* which Colombe had circulated in the first half of 1611. At that time Galileo explained to a friend that it was pointless to reply to someone who could not understand even the elements of astronomy (11: 152–3). That was perfectly sensible. Galileo did not mention that Colombe's tract ended by invoking the authority of the Bible, which contained many statements flatly opposed to Copernicanism. Colombe's point was that all theologians, without exception, say that when scripture can be understood literally, it should never be interpreted otherwise. He reinforced this by appealing to the authority of the Fathers of the Church. Colombe's tract could be disregarded as worthless, but this issue of the authority of scripture would have to be faced sooner or later (3: 290).

Galileo, indeed, had heard a disquieting rumour of an attack on Copernicanism in Florence on 2 November 1612; on the 5th a Dominican friar, Niccolò Lorini, assured him that he had not engaged in any philosophical dispute. He had merely chipped into a conversation to show he was alive, observing that the opinion of Ipernicus, or whatever he was called, seems to be contrary to scripture (11: 427). That false alarm confirmed how badly informed Galileo's critics were – and how jumpy he was.

It was Castelli who, more or less by accident, launched Galileo on an excursion into theology. It all started with a breakfast conversation at the court in Pisa about the satellites of Jupiter. The conversation went quite happily, but Castelli was called in afterwards by the Dowager Grand Duchess Cristina, who quizzed him about the compatibility of Copernicanism with scripture. Castelli was sure that she had been prompted by the philosopher, Boscaglia, for whom a moving Earth was unacceptable. Castelli's euphoric account, dated 14 December 1613, gave Galileo to understand that he had carried off the role of theologian with panache: he had satisfied the Grand Duke, the Archduchess, probably Cristina herself and in fact everyone present except the silent Boscaglia (11: 605–6). Whatever Galileo thought of this report, within a week he put his own reflections on paper in the form of a *Letter to Castelli*, which could be shown to friends (5: 279–88). Copies of this letter, some inaccurate, circulated widely. Its contents were later expanded into the *Letter to the Grand Duchess* of 1615. That pamphlet on the proper use of scripture in philosophical disputes was not published until 1636 and it may not have played much part in the dramatic events of 1615 and 1616. But it will be discussed in some detail below because it is essential to an understanding of Galileo.

Attacks and Delation to Rome

For most of 1614 there was little to worry Galileo. The Jesuits of the Roman College sided with him over floating bodies. Reliable reports that Grienberger was constrained by the General's insistence that Aristotle should be followed wherever possible were disappointing rather than alarming, since Galileo knew that Grienberger was well-disposed to him (12: 76–80, 112). Castelli was confident that the Tuscan court favoured Galileo rather than his opponents. But on 20 December a Dominican friar, Tommaso Caccini, used his turn as scriptural lecturer in Santa Maria Novella in Florence to denounce Copernicanism as contrary to scripture. Few commentators have much good to say about this friar. No doubt he saw himself as zealously defending the faith, but he did more damage than he could understand, let alone repair. It is true that in January 1615 Galileo received an entirely sympathetic letter of apology from a Dominican Preacher-General in Rome (12: 126–7) and, at the end of February, reassurances from Giovanni Ciampoli, another well-connected clerical friend there. But such a public attack was troubling, especially since Ciampoli passed on the friendly advice of Cardinal Barberini that Galileo should not go outside mathematics and physics: he should not provoke theologians by interpreting scripture (12: 145–6). This was little comfort to Galileo. He would have been very happy to confine himself to his own subjects, but what was he to do when others used theology to reject what he was saying? If no one qualified was going to defend the legitimate autonomy of astronomy and physics, then he would take the risk of doing so himself.

By this time Caccini's fellow-Dominican, Lorini, was no longer content to comment that what's-his-name, Ipernicus, seemed to contradict scripture. He thought the *Letter to Castelli* was a reply to Caccini. In February 1615, on behalf of his brethren at San Marco, he sent a copy to Cardinal Sfondrati, Secretary of the Roman Inquisition, with an informal covering letter. The accusation, oddly enough, was directed not against Galileo himself but against the 'Galileists', all good Christians, admittedly, but a bit too clever and obstinate in their opinions. Lorini describes them as showing off by spreading all sorts of 'impertinences' throughout Florence: his grievances about their interpretation of scripture or their attack on Aristotle certainly hit off Galileo's own views. Still, the only solid evidence was the *Letter to Castelli*. Surely, Lorini wrote, it is suspect and rash to say that scripture takes the last place in disputes about natural effects, that its expounders are often mistaken or that in natural things philosophical and astronomical argument counts for more than the sacred or divine. Lorini's copy of the *Letter to Castelli* was dangerously inaccurate in places, but his hostile letter did not distort Galileo's views as much as such documents sometimes do. He adroitly shielded himself from unpleasantness by asking that his covering letter should not be treated as a judicial deposition but kept secret as a loving service to his patron, the Cardinal (19: 297–8). He seems to have had his way, since he was never called to give evidence. The Inquisition very properly instructed its Florentine branch to acquire the original from Castelli, but we need not trace that part of their procedure since their own expert consultor eventually cleared the defective copy of the *Letter* of anything worse than unhappy expressions which could be misinterpreted (19: 305).

Meanwhile the irrepressible Caccini went to Rome and presented himself to the Inquisition on 19 March, secure in the conviction that his detailed information would enable the Church to save scripture from perversion. He recounted how in his lecture on Joshua he had modestly reproved Galileo's Copernicanism as contrary to scripture. The reasons he gave were, though he does not say so, identical with those in the concluding paragraph of Colombe's tract, with a quotation from a recent theologian borrowed word for word. Caccini had already explained to the Florentine Inquisitor what had prompted him to treat the topic, namely that some of Galileo's pupils were saying that God is not a substance, God is a sentient being and saints' miracles are not genuine. Lorini had later shown him the *Letter to Castelli* but, since the Inquisition already had a copy, Caccini contented himself with saying that it was public knowledge that Galileo held two propositions: one, the Earth moves as a whole, also with diurnal motion; two, the Sun is immobile. Both, Caccini thought, contradicted the faith. He added that some, including Lorini, suspected Galileo's orthodoxy because he was very friendly with the notorious Sarpi and corresponded with him still. The Academy Galileo belonged to was also in correspondence with Germans (meaning heretics), as could be seen from the *Letter on Sunspots* (19: 307–11).

Caccini's interrogators may not have worried much over Galileo's correspondence with Welser and Scheiner, though the charge of corresponding with Germans was included years later in the preamble to his condemnation. What they made of his friendship with Sarpi is not known, but it was no recommendation at that stage of the Counter-Reformation cold war. It took the Inquisition several months and a couple of very careful interviews at Florence with people named by Caccini to establish that his accusations against Galileo's pupils relied heavily on biased reporting and creative eavesdropping (19: 316–20). Towards the end of 1615 they were left with Galileo's Copernicanism, a query about the *Letters on Sunspots* and his interpretation of scripture in the *Letter to Castelli*, an interpretation which had not seriously disturbed their consultor. In the meantime Galileo had not been idle.

From the beginning of 1615 he devoted most of his energies to staving off a condemnation of Copernicanism. To this end he decided to expand his *Letter to Castelli* into a little treatise on how scriptural texts should be used in matters of science. By 6 January Castelli had engaged a sympathetic Barnabite priest to supply passages from Augustine and other Doctors of the Church in support of Galileo's interpretation of scripture (12: 126–7). On the 12th Cesi advised caution in reacting to Caccini's lecture: he had it from Bellarmine that Copernicanism was heretical. The less said about it at this juncture, wrote Cesi, the better: it could so easily be prohibited. The only safe ground for complaint was Caccini's wild attack on mathematics as a diabolical art (12: 129–30).

Although the proceedings of the Inquisition were conducted in the strictest secrecy, Galileo soon knew that Lorini had a copy, possibly inaccurate, of the *Letter to Castelli*, so on 16 February he took the precaution of sending a true copy to his friend, Piero Dini, in Rome. Dini was to seek the support of Galileo's 'very great friend' Grienberger and even, if possible, to show a copy to Bellarmine, whom his opponents wished to make their chief in the hope that Copernicanism would be condemned (5: 292). Bellarmine and Galileo can be seen as the protagonists in this dispute, so it is fortunate that we are in a position to compare their views in detail. Dini gave Bellarmine a copy of the *Letter to Castelli* and on 7 March reported a long conversation he had had with the Cardinal. Bellarmine had not heard any talk of Copernicanism. The worst that could happen to Copernicus's teaching was that it would be glossed as a merely hypothetical calculating device like Ptolemaic epicycles. Galileo could treat the topic in the same vein. The greatest scriptural hurdle to holding Copernicanism as true was the phrase in Psalm 19 (18 in the Vulgate numbering) about the Sun going forth as a giant: all interpreters treat this as attributing motion to the Sun. A reinterpretation of scripture should not be adopted hastily, but nor should it be hastily ruled out. Bellarmine concluded that he would

17   Robert Bellarmine, from Isaac Bullart, *Académie des Sciences et des Arts,*
tome 2, livre 1 (Amsterdam, 1682).

consult Grienberger. A quick check with Grienberger showed Dini that
the Jesuit mathematician would have preferred Galileo to have given
his proofs first and then argued for a reinterpretation of scripture.
Grienberger thought Galileo's arguments were plausible rather than
true (12: 151).

— 113 —

Bellarmine on Scripture

A letter in the same post told Galileo that he had an ally in Paolo Antonio Foscarini, a Carmelite friar who in January had published a book arguing that Copernicanism could be taken as compatible with scripture.[2] Foscarini sent a copy to Bellarmine, with the happy result that we have Bellarmine's assessment not only of Foscarini's opinion but also of Galileo's, since Bellarmine certainly intended Galileo to benefit from what he wrote. In his letter of 12 April Bellarmine courteously says that Foscarini and Galileo are prudent to content themselves with speaking only hypothetically, as Copernicus had always done. (It was not till later that Bellarmine realized that Copernicus had put forward his system as the true one.) There is no danger in saying that the hypotheses of the Earth's motions save the appearances better than do eccentrics and epicycles; nor does an astronomer need more. But to put forward heliocentrism as true runs the very great danger not only of annoying all philosophers and scholastic theologians but also of harming the faith by making scripture false. Foscarini may have shown that there are various ways of interpreting scripture but he would certainly find very great difficulty in interpreting all the passages he had quoted.

Bellarmine next reminds Foscarini that the Council of Trent forbade interpretations of scripture contrary to the consensus of the Fathers; in fact, the Fathers and all modern commentators take the passages about the Sun's motions literally. Bellarmine, we notice, did not consider properly whether this was just an unexamined assumption. Instead he goes on to his underlying theological conviction, nowadays referred to under its nineteenth-century label of *inerrancy*: there can be no errors in the Bible. It is no answer to say that heliocentrism is not a matter of faith, Bellarmine tells Foscarini: the topic itself may not be a matter of faith, but its source, the Holy Spirit, makes it one. It is just as heretical to deny that Abraham had two sons and Jacob twelve as to say that Christ was not born of a virgin.

Finally, Bellarmine does allow that if there were a conclusive proof of heliocentrism, then we would have to think again: no doubt he had consulted Grienberger on this point. But he will not believe there is such a proof until shown it. To show that heliocentrism saves all the appearances is not the same as proving heliocentrism true: it is very doubtful whether that could be proved and in case of doubt we ought not to abandon scripture as interpreted by the Fathers. We should also remember that the man who wrote 'The sun rises and sets and returns

to his place' was Solomon, who was not only inspired but the wisest
and most learned of men: he would hardly affirm something that
contradicted a proven or provable truth. It is no use saying that he
spoke according to appearances, the Sun appearing to move just as the
shore appears to someone in a boat to recede; the man in the boat
knows very well that the shore is fixed and corrects his error, whereas
no wise person needs to correct any error about the Sun and the Earth,
because he knows the Earth is fixed (12: 171–2).[3]

Galileo's Comments on Bellarmine

Galileo's own suggestions on how scripture should be interpreted will
be discussed in the next section, but it will be convenient to look first at
his (undated) reaction to Bellarmine's letter (5: 351–70). He thought
that those in authority were in danger of being misled on two important points. The first danger was that they would think a moving Earth
so paradoxical and absurd that it could never be demonstrated either
now or in the future. But many famous ancient philosophers had
thought it true, as did many moderns, including Copernicus, Gilbert
and Kepler, not to mention many Italians known to Galileo who had
never published their opinion. It was true that Copernicans were only
a small minority: that was because of the intrinsic difficulty of the topic.
But all Copernicans knew all the reasons for and against heliocentrism
and had had to overcome their own initial conviction that it was absurd, so their opinions should count for far more than those of people
who simply followed tradition unintelligently.

The second danger was that those in authority would believe that
Copernicus himself had proposed heliocentrism only hypothetically
and so would feel free to condemn it. But a reading of his *De
Revolutionibus* would leave no doubt that he put it forward as the true
system. The idea of its being only a hypothesis came from the preface,
which was clearly the work of someone else. Who would want to
condemn a book on the strength of a printer's or bookseller's blurb?
Confused talk about the role of epicycles and eccentrics in astronomy is
not helpful: there are certainly epicycles in the heavens, witness the
orbits of Jupiter's satellites. Ptolemy and Copernicus both took the
hypothesis of uniform circular motions realistically; Copernicus certainly took the hypotheses of the Earth's motions to describe what is
truly the case.

Galileo also answered the other points raised by Bellarmine: most of
them we can find in the *Letter to the Grand Duchess*. Philosophers and

theologians should not be irritated: if Copernicanism is false it can be refuted and condemned; if it is true, they should be pleased that others have opened the way to truth and saved them from condemning a truth. Catholic astronomers have no thought of making scripture false: some theologians should take care lest they make it false by interpreting it against propositions which may be true and demonstrated. It may be that Foscarini and Galileo will find it difficult to interpret scripture, but that is because of their ignorance, not because of any insuperable difficulty in reconciling scripture with demonstrated truths. Bellarmine is quite right not to accept without proof that the Earth can move: all Galileo asks is a thorough examination in which Copernicans should fail if they are not at least ninety per cent right. Copernicanism should not be thought absurd when most of what is said against it in astronomy and philosophy is false or worthless. That Copernicanism saves the appearances does not prove it true; but the received system cannot save the appearances, so it must be false, whereas Copernicanism could be true and, after all, no position can do more than save all the appearances. Bellarmine rejects the example of relative motion, where the shore seems to move away from the boat, because we are familiar with shores and boats and know the shore does not move. But the point of the example is simply to show that the Earth could be moving: if we could be transferred to the Sun and back, perhaps we would have similar knowledge of which is really moving. Galileo was quite genuine in his attempt to save the Church from a damaging mistake. Foscarini's *Letter* helped him to tackle the problem of reconciling Copernicanism with an acceptable reinterpretation of scripture.

Galileo's Interpretation of Scripture

In his *Letter to the Grand Duchess* Galileo did not see any obstacle in what the Council of Trent had said about the consensus of the Fathers; he made the sensible comment that even when a natural proposition in scripture is understood in the same way by all the Fathers, that interpretation is not binding unless the Fathers themselves actually discussed the matter thoroughly, going into both sides of the question and only then opting for one interpretation and rejecting the other. In other words, unexamined ideas, no matter how widely held, have no weight and the mobility of the Earth was never discussed adequately by the Fathers of the Church. In any case, as Galileo rightly said, the Council was concerned only about faith and morals, and Copernicanism raised

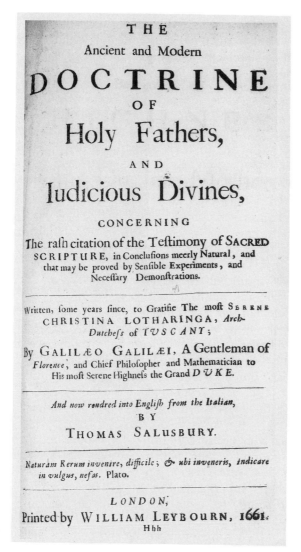

THE

Ancient and Modern

DOCTRINE

OF

Holy Fathers,

AND

Iudicious Divines,

CONCERNING

The rafh citation of the Teftimony of SACRED
SCRIPTURE, in Conclufions meerly Natural, and
that may be proved by Senfible Experiments, and
Neceffary Demonftrations.

Written, fome years fince, to Gratifie The moft SERENE
CHRISTINA LOTHARINGA, Arch-
Dutchefs of TUSCANY;

By GALILÆO GALILÆI, A Gentleman of
Florence, and Chief Philofopher and Mathematician to
His moft Serene Highnefs the Grand DUKE.

And now rendred into Englifh from the Italian,

BY

THOMAS SALUSBURY.

Naturam Rerum invenire, difficile; & ubi inveneris, indicare
in vulgus, nefas. Plato.

LONDON,
Printed by WILLIAM LEYBOURN, 1661.
Hbh

18   The title-page of Galileo's *Letter to the Grand Duchess* in Thomas
Salusbury's translation, *Mathematical Collections* (London, 1661).

no questions of morals, so he only had to show that to say the Earth is
a planet in no way involves matters of faith (5: 335–7). But it would be
a formidable task to show this, as can be seen from Bellarmine's letter
to Foscarini.

Galileo's problem, and Bellarmine's too, was that Copernicanism clearly contradicted the literal meaning of many statements in the Bible. Everything in the Bible, they agreed, is inspired, with the consequence that no statement in it can be called erroneous. Galileo's problem was how to convince Bellarmine that Copernicanism could be accepted without imputing error to its human authors and so to its Divine Author.

Bellarmine, for his part, is frequently praised, and not only by Catholic apologists, for advice which was appropriate, so it is claimed, not only to the specific issue of Copernicanism in 1615, but also to any similar clash of theological and secular ideas. He is praised first because he allowed to Foscarini that, if Copernicanism could be proved to be true, then we would have to say that we do not understand the geostatic passages of scripture. So he is held to have had an open mind and merely to have made the reasonable stipulation that he would insist on the received understanding of scripture until proof was produced that it was untenable. It is difficult to know quite what to make of this: what would Bellarmine have accepted as an adequate proof? Kepler had already published his vastly improved new astronomy, though it had not yet made much impression. It is very likely that Bellarmine, if asked, would have been happy to welcome it as an improved set of calculating devices, but would have wanted more before taking seriously the claim that the Earth really does move. That does not make him a fictionalist or instrumentalist or conventionalist philosopher of science, except in planetary theory. In general, he was a realist; in particular, it is not unfair to say that he *knew* the Earth was fixed.

But those who praise Bellarmine for espousing a philosophy of science more adequate than Galileo's are also impressed by the serenity with which the Cardinal was happy to see Copernicanism put forward as a hypothesis superior to Ptolemy's. He would not accept it as proven, and rightly so, for it was not proven, but (so this appraisal goes) he had no objection to its being held as a hypothesis. So Galileo was clearly in the wrong for expecting and perhaps demanding more.

Bellarmine, as Galileo recognized, was obviously in the tradition succinctly and notoriously expressed in Osiander's preface to the *De Revolutionibus*: astronomical hypotheses are merely calculating devices; they need have nothing to do with questions of truth or falsity; they may even be known to be false, yet still usable, the hypotheses of the Earth's motions being a case in point. Now it is perfectly understandable that Bellarmine should have expected astronomers to remain content within the confines of this instrumentalist approach to astronomy. As we have seen, he believed that Copernicus himself had required no more. But this was a far cry from tolerating Copernicanism as a hypothesis in its usual modern meaning: something which is as yet admittedly unproven, but which might one day be shown to be true, or at least nearer the truth than other hypotheses. A calculating device is not a guess which becomes more credible with increasing evidence: it is not that kind of thing.

So Galileo was not being pig-headed when he refused to rest content with the proposed compromise of treating Copernicanism hypothetically in Bellarmine's sense, since he knew that this meant accepting that the Earth is in fact immobile. It may be suggested that he should have pretended to accept the compromise and under this cover further the Copernican cause. Yet that, in fact, was what he tried to do later in the *Dialogue* for which he was condemned: the only allowable alternative to conclusive proof was a mere calculating device. Once one realizes what Bellarmine (and the Congregation of the Index in 1616 and 1620) meant by 'hypothesis', then the whole strand of criticism which praises Bellarmine's philosophy of science to the detriment of Galileo's begins to look pretty frayed. Bellarmine, of course, was faced with the hard task of integrating stunning novelties into his world-view. Moreover, the philosophical question of the status of scientific hypotheses is still highly controversial, so it is not surprising if he could not answer it satisfactorily: neither could Galileo.

Galileo, in fact, had a good deal in common with Bellarmine and other Catholic theologians. They agreed that two truths could not contradict each other: there would be no fudging the issue by claiming that something could be true in philosophy but false in theology. Galileo is also able to quote from Pereira's commentary on Genesis the following very sane principle or rule of interpretation:

> In dealing with what Moses taught we must beware of, in fact, avoid altogether, thinking and asserting as a positive fact anything which goes against the manifest experiences and reasonings of philosophy or other disciplines, for since every truth always agrees with truth, the truth of Holy Scripture cannot be contrary to the true reasonings and experiences of human teaching.

So Galileo here had an influential Jesuit amplifying an echo from St Augustine which Galileo goes on to quote.[4] It meant that Copernicanism could be made theologically acceptable by bringing forward conclusive reasoning (demonstrations) and observations, something which, in principle, Pereira, Bellarmine and all other theologians would allow. Yet such agreement on principles of interpretation settled little. Pereira himself used it to show that people like Lactantius, who rejected the Antipodes on scriptural grounds, had been quite mistaken, a point Copernicus had also made in his book. But, Pereira thought, Chrysostom had been equally mistaken to think that the stars move through the heavens like fish in water, presumably because contemporary Aristotelian science claimed to show by philosophical reasoning that the stars and planets must be attached to spheres. (For all their talk of demonstration and proof Aristotelians were not very exacting in their actual standards: lots of things escaped scrutiny because they were such familiar furniture in the home inherited from Aristotle.) The anti-Aristotelian idea of stars moving freely was one which Bellarmine took over from Chrysostom and other Church Fathers: he espoused it in his Louvain lectures of 1570–2 and never abandoned it, though he did not publish it, as it was in contrast to the general Jesuit policy of adhering to Thomistic Aristotelianism as far as possible. So, though Galileo, Bellarmine and Pereira all accepted the principle that scientific and theological truths can never conflict, along with the principle that established scientific truths can help us to understand scripture, they were still a long way from a meeting of minds. It must also be remembered that, unlike Galileo, theologians had a professional suspicion of novelty, all the deeper at this period of the Counter-Reformation. Pereira is quite sharp about Cajetan, who in his commentary on Genesis had pleaded for more elbow-room to accommodate orthodox new ideas, on the enlightened ground that otherwise interpreting scripture would be just like copying from a book into a notebook.[5] There is, of course, a conservatism essential to any theology which claims to pass on an original divine revelation intact and undistorted, but it brings with it the hazard of treating what once had been novelties, such as Aristotelian cosmology, as inextricably bound up with the revelation contained in scripture. (No doubt it helped if one could forget that Aristotelian ideas and methods had first met with great resistance from many theologians and that their tenancy of Christian philosophical quarters might seem statutory but in fact rested merely on their customary acceptance during the preceding three centuries or so.) Bellarmine was less inclined than Pereira to squeeze scripture into an Aristotelian mould, but he was at least as committed to insisting that the Earth is fixed, as scripture so plainly gave him to understand.

Galileo was clearly calling for a major shift in theology, something which could be as momentous for theology as the shift in astronomy and physics he was helping to bring about. Bellarmine cannot be blamed if he was apprehensive that the whole basis of theology, indeed of faith, might be eroded, as indeed it has been for so many people who came after him. Galileo knew he had to allay these fears: he knew that he had to get people to see that there could be a new way of looking at things, admittedly different from the old, yet with some support and encouragement from within traditional theology, a way in which nothing essential would be lost and much might be gained. Later we shall see him in his *Dialogue* trying to persuade the reader that there could be a new way of looking at the familiar facts of falling bodies and projectiles, so that even on a moving Earth bodies would fall to the foot of a tower or of a ship's mast. What new way of reading the geostatic passages of scripture could he advocate that would persuade people that Copernicanism was compatible with them?

Well, he was quite firm that the Bible cannot err, provided that its true meaning is understood. He did not need to spend long on the fact that not every statement in scripture can be taken literally, otherwise we would be landed with the heretical notion that God has hands and feet and gets angry and forgets (a point which Caccini had garbled). To call such language metaphorical was the merest commonplace, as can be seen, for instance, in Bellarmine's *Ladder, wherby our mindes may ascend to God, by the stepps of his creatures*, first published in that very year. Bellarmine mentions that God measures the earth with three fingers and the heavens with a span and continues quite comfortably: 'But this is spoken metaphorically, for *God is a Spirit*.'[6] Yet, as Galileo well knew, Bellarmine could also be very literal-minded.

Still, the fact that scripture clearly contains metaphorical language gave Galileo the opportunity to lay down a key principle: such propositions, dictated by the Holy Spirit, were put forward in that way by the sacred writers to accommodate them to the capacities of the rude and unlettered common people (5: 315). This principle, he says, is so generally accepted by theologians that it needs no evidence, though he does back it up later with well-chosen quotations. We might call this the principle of accommodation, meaning that God kindly accommodates the expression of what he wants to say to suit the limitations of people who think 3 is a fair enough value for π. But 'accommodation' already had a different traditional meaning: it referred to the practice of applying, and often misapplying, scriptural texts in a meaning other than the original. An instance would be the inscription on a bust of Lorenzo the Magnificent which greeted Leo X on his entry into Florence: 'this is my beloved son.'[7] A satisfactory traditional name for it is the principle of condescension: God condescends to human limitations and talks in a language people can cope with. Galileo's claim is that scripture refers to astronomy only incidentally: whenever it talks casually of Earth, water, the air, the Sun or any created thing, then the principle of condescension has an application.

It occasionally puzzles modern readers to find Galileo naïvely attributing to the biblical authors a knowledge of the true system of the universe, so that they are partners in the divine condescension. But in this he is simply following St Augustine who, at his most lordly and dismissive, determined a discussion on the shape of the universe thus:[8] 'In short, our writers knew the truth about the form of the heavens, but the Holy Spirit who was speaking through them did not wish to teach men those things that are useful to no one for salvation.' This contained what Galileo needed: even if it had occurred to him, there would have been no point in challenging the common view, emphasized by Bellarmine, that Solomon was the wisest man who ever lived, or Pereira's belief that Abraham was an expert astronomer and Moses was the only expert on creation. Augustine was very useful to Galileo: he advised that in matters that are obscure we should believe nothing rashly, lest through love of our own error we might come to hate something which truth later reveals to be in no way contrary to the holy books of the Old or New Testament.[9] So there could be new interpretations of scripture, even ones which would allow the truth of Copernicanism, provided that Galileo could show there was good reason to take Copernicanism sufficiently seriously to make a reinterpretation of scripture legitimate.

Now Galileo himself certainly thought that the true system of the universe can be discovered, so he was not going to be content with mere calculating devices. He clearly thought that this system was none other than Copernicanism, not indeed the major revision of Copernicanism put forward by Kepler, but Copernicus's own system as improved by Galileo. But he had no conclusive proof that this was the true system, nor did he ever find one. By the time that he wrote his *Dialogue* he seems to have thought, mistakenly, that he could prove Copernicanism, especially by his explanation of the tides: indeed, even in early 1616 he was pinning his hopes on this argument. In 1615 he certainly claimed, reasonably enough, to be able to refute the arguments of Ptolemy and Aristotle. His telescopic discoveries fitted Copernicanism admirably and plainly confuted the Ptolemaic system (5: 311). What he did not do at any time was show the inadequacies of the Tychonic compromise, which kept the Earth fixed but could accommodate all the new discoveries. He preferred to rely on the claim that he could produce arguments about natural effects which could be explained in no other way than by making the Earth move (again, he was thinking of the tides).

By 1615 Galileo was primarily concerned to prevent an ill-judged condemnation of Copernicanism: he was not saying that it should not be condemned, but that it should not be condemned without being properly understood. It is perfectly true that he was over-confident and that he had been campaigning for some time to get the new system accepted. But his main preoccupation was to avoid a premature intervention by Church authority, a point he developed at some length. He had certainly reflected on the implications of what was at stake. He contrasted physical propositions about which we can only make conjectures, for instance, whether stars have souls, with others 'of which we have, or can firmly believe we could have, undoubted certainty, through experience, long observations and conclusive demonstrations'. In other words, in natural questions which are not matters of faith, we should start by asking whether anything is, or is the sort of thing that could be, proved beyond doubt or known through sense experience. The warning is clear enough: even an unproven set of statements may come to be proved later (5: 313, 330, 332, 338–9).

All commentators on this letter point out that Galileo more than once undermines his own carefully constructed case. He starts from the non-controversial position (shared by Pereira and Bellarmine) that theologians have to show that proven secular knowledge is compatible with scripture. But if a proposition of secular knowledge is not proved conclusively and contains something contrary to scripture then, says Galileo, 'it is to be reckoned undoubtedly false and shown to be so by every possible means' (5: 327). Now, if Galileo had been able to demonstrate the truth of Copernicanism, he would have been quite safe on this high ground. Since he could not provide such a proof, it followed from his own explicit rule that theologians should reckon Copernicanism was false and should make every effort to prove it false. It might also seem to follow that Copernicanism could be condemned without further ado.

Galileo had undermined his own position. It must also be said that his rule is seriously misconceived. It could never be sensible to say that any scientific statement which conflicts with the received understanding of scripture and which has not been proven conclusively should be treated as false by theologians. After all, if such conflicts occur it will be precisely in the case of scientific statements which, though not proven, seem to have so much in their favour that their proponents are hopeful or even confident that they will eventually be established. That was the status of Copernicanism in 1615. The sensible thing to say in such cases is 'wait and see.' The fact that Galileo's suggested rule is indefensible means, of course, that it cannot be used to justify the condemnation of Copernicanism, even if Galileo had sawn off the branch he was sitting on. It is also worth noting how it was that Galileo committed himself to this indefensible rule.

He was trying to stave off a condemnation. He was confident that if the issue was discussed fully Copernicanism would not be condemned. He was over-confident that he could prove the system true; in his more sober moments he was content to claim that he could make out for it a much better case than could be made out for its rivals, including Tycho's compromise system. So why did he talk about necessary demonstrations and sensible experiences? Partly because this is how he himself saw the goal of science, but also because in all the tradition going back from Bellarmine and Pereira to Augustine these were the only things that offered a chance of convincing theologians that scripture needed a novel interpretation. Nothing less than a demonstration would make Bellarmine rethink scriptural interpretation. Nor could Galileo get round the fact that Augustine himself had proposed something very like the rule he unwisely espoused.[10] As McMullin brings out clearly, the tensions and inconsistencies in Galileo's exposition are only a more pronounced version of those in Augustine's.[11] Bellarmine was emphasizing the strand of Augustine's thought which called for scientific proof prior to taking scripture in anything but the literal sense, whereas Galileo made more of the strand which made scripture seem irrelevant to details of natural science. If Galileo had confined himself to showing that scripture was irrelevant to the detailed findings of astronomers, he would have been less vulnerable. No doubt he felt he had to show that the miracle of Joshua was better explained in Copernicanism; it had, after all, been a principal issue in Castelli's conversation with the Grand Duchess. No doubt he exaggerated the cogency of his own proofs of Copernicanism. His arguments may seem a bit of a mixture but his position, I think, was this: if an immediate condemnation could be averted, then in the breathing-space obtained full discussion would show that Copernicanism was at least the best available system and one which was getting nearer all the time to being proved; hence it could be tolerated as probable, that is, as a legitimate opinion to hold and to support as the kind of thing that could eventually be shown to be true.

Another aspect of the matter that is usually overlooked is that Galileo was, I think, trying to offer reassurance to people who were understandably worried that scripture might be left undefended from bizarre and subversive private opinions. If the interpretation of scripture on something as fundamental as the structure of the created universe could change, would theology not be exposed to the challenges of all sorts of extravagances in the name of science? Not so, according to Galileo's rule: treat all such opinions as false and show that they are false, unless they are demonstrated. This advice was firmly based on Augustine, but it was unwise, especially since it undermined Galileo's own position. But it was not the advice of a subversive thinker, or of a zealot who conjured up the over-simplified alternative: either condemn Copernicanism completely or accept it outright.

Yet Galileo is frequently criticized for having tried to shift the burden of proof onto theologians and Church authorities. But to say that *philosophers* should leave scripture alone and try to prove him wrong by reason and observation was hardly shifting the burden of proof. He did say that before a natural proposition can be condemned (meaning by theologians or Church authorities) it must be shown that it has not been demonstrated necessarily (5: 327). But that is sound advice: it comes down to saying 'do not condemn a proposition without first ascertaining that it has not already been proved true.' In his own case it would not have been difficult for Bellarmine (with Grienberger's help) to show that Copernicanism had not been proved true. Galileo also said explicitly that a proposition is not contrary to the faith until it has been refuted by most certain truth. This *does* sound like a shifting of the burden of proof; it is, however, a perfectly fair quotation from Augustine.[12] But there is one very outspoken passage:

> It is not in the power of any created being to make propositions true or false differently from what is *de facto* the case in nature. So it is more advisable to ascertain first the necessary and unchangeable truth of the fact, which no one can overrule, than without such assurance to condemn one side . . . In brief, if it is not possible that a conclusion should be declared heretical while there is doubt that it may be true, then the endeavour of those who claim to damn the earth's mobility and the sun's stability must be futile, unless they first show that they are impossible or false (5: 434).

Galileo had not managed to produce the one thing that would move theologians or Church authorities to allow a novel interpretation of scripture, namely proof positive that Copernicanism was true. He was trying to gain time for a full and impartial discussion of the issues: the one thing he wished to avert was the kind of casual and routine intervention that actually took place in 1616. But the fact remains that in the passage just quoted he did try to shift the burden of proof. That he did so is one of the main counts against him in Koestler's attempt to make him the originator of modern Western cultural schizophrenia.[13] It is excused as contrary to Galileo's real intentions by the sympathetic Langford.[14] I suspect that Galileo knew what he was doing and intended that the burden of proof should be shifted.

He was talking only about scientific statements. His position was that science cannot be derived from scripture, but only from reason and observation, though it has to be compatible with scripture rightly understood. So the established propositions of science should be used to interpret those passages in scripture which allude to natural things: if necessary, scripture should be reinterpreted to fit in with what is known from science. This position, incidentally, is nearer than Bellarmine's to what is nowadays the officially approved theory and practice of the Catholic Church. I take Galileo to be saying: where there is a conflict between the received interpretation of scripture and some part of science, the only ground that theologians or Church authorities can have for condemning a scientific proposition is its falsity and its falsity can be shown only by scientific arguments. This would not necessarily make theology unduly dependent on science or on views put forward in any non-theological discipline. It would, for instance, leave room for warnings, short of condemnations, against scientific views whose compatibility with faith was not clear. It would certainly face theologians with difficult tasks, much more difficult than Galileo could have envisaged. But whatever the shortcomings of his proposals, he did see that some such tasks had to be faced and he did his best to help theologians.

Galileo in Rome

It is not known when Galileo completed the *Letter to the Grand Duchess* but it certainly represents his opinions at the close of 1615. It was also written in deliberate disregard of the friendly advice he had received in the course of the year from influential and knowledgeable supporters and from at least three cardinals to keep out of the sacristy and confine himself to mathematics and physics. We have seen why he could not be content with treating Copernicanism 'hypothetically'. In May he explained his position to Dini. If he treated Copernicanism realistically anyone could say it was against the faith; if he then tried to show it was not, he was told not to deal with scripture. The best way to show Copernicanism was compatible with scripture would be to produce a thousand proofs of its truth, not an easy thing to do when his opponents are so ignorant. There would be hope if he could come to Rome and explain things orally. Given recovered health, that was what he hoped to do in order to save the Church from making a mistake. All this he wrote to Dini in May (12: 183–5). By 11 December he had arrived in Rome.

He came with the recommendation of the Grand Duke and was entrusted to the care of a very disgruntled Florentine ambassador, who was sure no good could come of the visit. Galileo's triumphal visit four years earlier had ended none too soon when Bellarmine had been on the point of calling Galileo to account. Now some Dominicans were ill-disposed to Galileo. This was certainly not the time or place to defend new teachings (12: 207).

Galileo, characteristically, was optimistic that he would clear his name and defend the cause of Copernicanism. He began what he expected to be a long round of visits to cardinals and other dignitaries. He also gave virtuoso displays of his formidable argumentative powers, just the very thing that the ambassador had dreaded as likely to lead to disaster. The secrecy of the Inquisition did not prevent Galileo from learning that the matter had been under discussion for months and that Lorini had had some part in it. But the same secrecy did bar Galileo from direct contact with those who mattered, since he was supposed not to know anything of what was going on (12: 223, 225, 227).

By 6 February 1616 he reported that his own name was cleared but that he needed to stay to defend Copernicanism. In the same letter he recounted a strange visit from the brazen Caccini who tried to dissuade Galileo of what he knew for certain, namely Caccini's delation (12: 230–2). This seemed to Galileo a good time to ask Cardinal Orsini to talk to the Pope in favour of Copernicanism: Galileo's pet argument that the tides implied a moving Earth would surely win round Paul V. Orsini obliged enthusiastically shortly afterwards. Unfortunately he succeeded only in provoking the Pope to refer the matter to the Inquisition (12: 242).

On 19 February the Inquisition's consultors were asked for their opinions on two propositions:

(1) The Sun is the centre of the world and consequently immobile with local motion.
(2) The Earth is not the centre of the world nor immobile, but moves as a whole, also with diurnal motion.

The awkward English reflects the original Italian, which in turn was derived from Caccini's delation. No doubt some Inquisitor could have come up with something less clumsy, if the Inquisition had thought it was worth the trouble. The consultors had their judgement ready on the 23rd. All agreed that the first proposition was foolish and absurd in philosophy and formally heretical in that it expressly contradicted many sentences of scripture, taken in the proper meaning of the words and according to the common exposition of the Fathers and learned

19   Paul V, from Alphonso Ciaconio, *Vitae et res gestae pontificum Romanorum*, tome 2 (Rome, 1630).

theologians. All gave the second proposition the same censure in philosophy; theologically it was at least erroneous in faith (19: 320–1). The consultors, of course, could only advise, not act. The Inquisition chose to act formally on this occasion by getting the Congregation of the Index (the department in charge of censoring publications) to publish a

Friderici Achillis Ducis Wirtemberg. Confultatio de Principatu inter Provincia
:uropæ, habita Tubingę in Illuftri Collegio , anno Chrifti 1613.
Donelli Enucleati , *five* Commentariorum Hugonis Donelli de Iure Civili in com
endium ita redactorum, *&c.*
Et quia etiam ad notitiam præfatæ Sacræ Congregationis pervenit, falfam illar
octrinam Pythagoricam , divinæque Scripturæ omnino adverfantem de mobilitat
'erræ,& immobilitate Solis , quam Nicolaus Copernicus de revolutionibus orbiur
æleftium , & Didacus Aftunica in Iob etiam docent, jam divulgari,& à multis re
ipi, ficuti videre'eft ex quadam epiftola impreffa cuiufdam Patris Carmelitæ , cu
tulus ; *Lettera del Reu. Padre Maeftro Paolo Antonio Fofcarini Carmelitano fopra l'opinio
e de' Pittagorici , e del Copernico , della mobilità della Terra , e ftabilità del Sole ; & Il nuo
o Pittagorico Siftema del Mondo, in Napoli per Lazzaro Scoriggio* 1615. in qua dictus Pate
ftendere conatur,præfatam doctrinam de immobilitate folis in centro Mundi,& mo
ilitate Terræ,confonam effe veritati , & non adverfari Sacræ Scripturæ: Ideò ne vl
:rius huiufmodi opinio in perniciem Catholicæ veritatis ferpat , cenfuit dictos, Ni
)laum Copernicum de Revolutionibus orbium , & Didacum Aftunica in Iob fu
endendos effe donec corrigantur. Librum verò Patris Pauli Antonij Fofcarin
armelitæ, omnino prohibendum,atque damnandum, aliofque omnes Libros parite
em docentes, prohibendos , prove præfenti Decreto omnes refpectivè prohibet
amnat,atque fufpendit. In quorum fidem præfens Decretum manu, & figillo Illu
riffimi, & Reverendiffimi D. Cardinalis Sanctę Cæciliæ Epifcopi Albanen. figna
m,& munitum fuit , *die* 5. *Martij* 1616.
P. Epif. Albanen. Card. Sanct. Cæciliæ.

Locus † figilli.

Regift. fol. 90.

20 The Index's ruling of 1616, from *Index librorum prohibitorum* (Rome, 1667).

decree on 5 March. This decree was meant to put an end to the spread
of the false doctrine of the mobility of the Earth and the immobility of
the Sun. Foscarini's book, which argued that Copernicanism was compatible with scripture, was consequently prohibited, as were all other
books teaching the same thing. Copernicus's book was merely suspended until corrected (19: 322–3). This meant that it could be read
freely once the necessary changes had been made to make it read
'hypothetically': as a set of calculating devices it was very useful to
astronomers and no danger to anyone else.

The decree made no mention of Galileo, though his *Letters on Sunspots* could easily have been corrected. In fact, he had already been
dealt with, privately but officially, by Bellarmine on the express order
of the Pope. The intention was to spare Galileo and the Grand Duke
humiliation. But the Pope's order of 25 February was severely phrased.

Bellarmine was to advise Galileo to abandon Copernicanism. If he refused then the Commissary, before a notary and witnesses, was to give him an order to abstain completely from teaching or defending or treating of this teaching and opinion; if he did not comply he was to be imprisoned (19: 321).

The efficient Bellarmine called in Galileo the next day. What actually happened on that occasion is the subject of insoluble dispute. No one doubts that Galileo, like every other Catholic, was bound by the terms of the Index's decree. Bellarmine certainly advised him that Copernicus's doctrine could not be defended or held. On 26 May he obligingly gave Galileo a certificate confirming this (19: 348). Not surprisingly, Galileo's recollection was that nothing else of significance had taken place; it was in all conscience a disastrous enough setback to all his hopes but as far he knew, like any Catholic, he could still treat Copernicanism 'hypothetically'. The problem centres on a document which surfaced at his later trial and played an important role in his condemnation. This unsigned account of the interview with Bellarmine has the Commissary of the Inquisition giving Galileo a very solemn injunction to relinquish Copernicanism altogether and not to hold, teach or defend it in any way, verbally or in writing. Galileo, according to this minute, promised to obey (19: 321–2). Clearly, to break such a promise would be a much more serious offence than would a contravention of the milder general regulation, since the injunction singled out Galileo for special treatment. Its actual wording might be construed to allow him to discuss Copernicanism 'hypothetically', or at least to use it in calculations, but the intention seems to have been to silence him completely. One does not have to think well of this procedure to see that it may have been the only effective one for the ecclesiastical authorities to use: given the slightest opening for discussion Galileo would have had Copernicanism back on the agenda sooner or later. Was he served such an injunction? No one knows.

The minute is not only not notarized, it is not even signed, yet it does not seem to be a forgery. The most likely explanation is that it is the product of bureaucracy at its scruffiest, but nevertheless a true account of what happened in Bellarmine's house. That a subdued Galileo should genuinely have no recollection of it is adequately accounted for by Bellarmine's certificate, which made no mention of it, though one has to conclude that Bellarmine himself did not see Galileo as being more constrained than any other Catholic. (Bellarmine himself would have interpreted the public decree strictly so he may have thought it sufficient to clip Galileo's wings; he may, in other words, have understood the personal injunction as merely spelling out the fact that

Copernicanism could be no more than a calculating device. This may not sound very convincing: alternative interpretations of what happened are no more convincing and a good deal more conjectural.) If the Inquisition had allowed proper defence counsel to the accused and not protected itself so thoroughly by secrecy, a competent lawyer at Galileo's later trial might well have been able to get the injunction thrown out. As it was, it played a large role in his condemnation.

There are, then, extra grounds for sympathy with Galileo at his trial. If he had ever been served the injunction and promised to obey, Bellarmine's certificate was assurance enough that he was no more constrained on this topic than any other Catholic. (Bellarmine, unfortunately, died in 1621, long before the trial.) On that point he was in good faith. It is, however, important to notice that even had no personal injunction ever been given him, the Inquisition would still have had sufficient grounds to condemn his *Dialogue* (see chapter 8). I say this not to defend the Inquisitors' actions, though they have as much claim to a fair hearing as Galileo, but to bring out the fact that it was the intervention of the Inquisition and the Index in 1616 that was at the root of the trouble. That was when it all went wrong, even though Bellarmine's compromise of tolerating Copernicanism 'hypothetically' no doubt seemed to him and his fellow-Inquisitors to be sufficiently liberal. Galileo's attempts to expose the inadequacy of that compromise had failed: even to this day he is often blamed for provoking the official intervention of the Church. But Galileo realized the shortcomings of the compromise. Whatever his exaggerations or inconsistencies, he will always be a potent symbol of independent and constructive thought: he asked better questions and gave better answers. He was not given the opportunity he desired to save his Church from a damaging mistake. It is not anachronistic to wonder why he could not have been given at least as much official attention as Caccini was. It is, however, pleasing to record a tribute from Pope John Paul II: 'Galileo formulated important epistemological norms indispensable in reconciling Holy Scripture and science.' We shall see in chapter 10 that the Pope's remark heralded a re-examination of the Galileo affair.[15]

7

Controversy and
New Hope

By June 1616 even Galileo had to accept that there was no further point in his remaining in Rome. He returned to Florence with his good name intact. In addition to Bellarmine's certificate he had commendations from cardinals to the Grand Duke and he had had a friendly audience with the Pope (12: 263–4). He soon learned that he had not lost the sympathy of the Jesuits. Grienberger and Guldin visited Cesi in September. He found them well-disposed to Galileo and both, especially Guldin, displeased at the way things had turned out (12: 285). Although his campaign had failed, Galileo was not in the least inclined to abandon Copernicanism; he would simply await better days. In the meantime there were still ways of winning intellectual glory. Even before leaving Rome he had returned to a problem which was of vital interest to naval powers: how to find longitude at sea (12: 256). Without reliable timekeepers sailors could not find the difference between their time and the time at a given place ashore; in other words they could not find their longitude. (In old age Galileo did sketch how a pendulum could be employed in clocks, but not even the pendulum-clock constructed independently by Huygens would have served: a sufficiently accurate marine chronometer was not to be made for another century and a half.) Galileo's solution, which he had hit on in 1612, was to use the natural clock he had discovered, namely Jupiter's satellites; the eclipses of the satellites would be observable much more frequently and accurately than the lunar eclipses which navigators had had to rely on hitherto (5: 419–24). With accurate tables for satellite positions visible from a given station ashore at a given time, an observer at sea would be able to find the required difference in time. In 1612 Spain had shown no interest in Galileo's first approach.

Galileo knew that his method was theoretically sound. He had the support of the Grand Duke, who engaged the Tuscan ambassador to give all the necessary help in dealing with Spanish officials. He himself was ingenious in thinking up ways of lessening the difficulties of using a telescope to make tricky observations on the high seas. He foresaw the need to present the tables in a form that would not flummox mariners. His letters on the topic are models of advertising, since he himself was fully persuaded that the method could be made to work. He was even prepared to go to Spain to train officers in the method, though perhaps that offer was partly inspired by disgust over the way he had fared in Rome (12: 291–5). In any event, despite all the ingenuity he put into it this project never came to much. But he never gave it up. In his old age he was engaged in serious negotiations with another maritime power, the Dutch. What impresses us now is not so much the invincible optimism of Galileo as his untiring observations of the satellites. By early 1618 he had completed corrected tables, called after Bellosguardo, where he rented a villa from November 1617. It lives up to its name, with a splendid view of Florence in the valley below.

Although Galileo could not advocate Copernicanism openly, he could place an unpublishable manuscript with a trustworthy prince who would know what he was up to. So in May 1618 he sent Leopold of Austria the essay on the tides which Cardinal Orsini had tried to draw to the Pope's attention two years earlier (5: 377–95): naturally Galileo had to characterize it as 'a dream' and 'a caprice'. Nothing, however, came of this beyond a polite reply (12: 390–2, 397–8).

Comets

In late November 1618 Galileo, confined to bed by illness, was unable to observe the last and most remarkable of three comets that appeared from August onwards. Nevertheless he was widely looked to for clarification of the nature of these puzzling phenomena, yet he had not openly published his thoughts about the new star of 1604 because he had not known what to make of it. His decisive and not entirely critical adoption of Copernicanism was accompanied by a much more cautious, almost sceptical, assessment of what could be said reliably about new stars and comets. This caution, noticeable also in his first letter on sunspots, is characteristic of a significant part of his work; it can be disconcerting because there is a temptation to think that his less successful ideas must come from the sort of swooping flights which pounced on his greatest discoveries. That is far too tidy. Some of his successes were backed by years of hard work and cautious sifting of interim theories; some of his refusals to accept ideas that seem to us more promising than his own were not unreasonable, though we may detect something of a partisan desire to make a case (or spoil another's). This is a contentious topic, so it need not surprise us if his contemporaries were sometimes puzzled and could not take his doubts and tentative conjectures at their face value.

Galileo soon entered public discussion of the comets, gradually became entangled in increasingly testy controversy, and extricated himself by a masterpiece of polemic, which delighted his supporters but irritated and even alienated some who hitherto had been well-disposed or at least not overtly hostile to him. In particular, his masterly attack in *The Assayer* of 1623 on Orazio Grassi, professor of mathematics at the Roman College, may well have lost him influential friends: he had already alienated Scheiner earlier in the controversy.

It all started quietly with the publication of an astronomical disputation at the Roman College (6: 23–34), a routine enough occasion since academic institutions were expected to stage special lectures to discuss or celebrate unusual events. The author was given simply as 'one of the fathers of the Society'. The unnamed Grassi did not attack Galileo: if anything, he saw himself as helping him indirectly by showing that the telescope was a reliable and useful instrument which had already revealed many new things and might be of use in settling disputes about comets (6: 25, 33). So far from following Aristotle in placing the comet below the Moon, Grassi used observations of parallax to show it was between the Moon and the Sun; he even supported this by a well-meant, though confused argument, which was probably based on a misunderstanding of Galileo's work: the further an object is away from us, the less it is magnified by the telescope.

A letter of 2 March 1619 informed Galileo of this publication and added the remark that some people in Rome, not Jesuits, were saying that it overthrew the system of Copernicus (12: 443). Early in June Mario Guiducci, Consul of the Florentine Academy, published a *Discourse on the Comets* in which he included extensive reports of Galileo's conjectures (6: 39–105). Guiducci had been educated at the Roman College and was also a pupil of Galileo's. The bulk of this work should certainly be treated as Galileo's own, though literary conventions may well have allowed him to insist that it was his pupil's. Grassi quite reasonably treated Guiducci as Galileo's spokesman. I shall do the same: their later protests may be looked on as legitimate controversial ploys of minor importance.

The main point of the *Discourse on Comets* is that observations of parallax are irrelevant to fixing the location of comets unless we can be sure that they are real permanent objects. There are serious grounds for wondering whether they have not more in common with phenomena like rainbows or the projection of the Sun's rays through broken clouds (6: 71–3). This is not just a cavil to make a difficulty: all explanations are doubtful, including Galileo's that they could be reflections from vapours. These arguments are directed specifically against Tycho Brahe, who wrote about the comet of 1577, and against the mathematical professor at the Roman College, who seems to subscribe to everything Tycho said (6: 64–5). The upshot is that the location of the comet has not been established. Neither has its motion, which could be rectilinear (6: 90–1). Incidentally, the way the telescope works is misunderstood at the Roman College (6: 94). The Aristotelian contrast between heaven and Earth is empty: as far as that goes, comets could be matter from Earth gone up to immense heights: anyway, we know from sunspots that there can be smoky vapours in the heavens (6: 93–4).

On 12 August Ciampoli reported from Rome that the Jesuits were offended and were preparing a reply (12: 466). Another friend, writing from Vienna on 24 August, revealed that the anonymous Jesuit author was Grassi. He also reported that Scheiner intended to pay Galileo back in the same coin (12: 487–9). (He was entitled to: the early part of the *Discourse* may be more Guiducci's responsibility than the rest of the work, but it went out of its way to insult Scheiner as incompetent and smear him as one of the those who had plagiarized Galileo's discoveries (6: 48).) Grassi did not take long to produce his reply, *The Astronomical Balance* (*Libra astronomica ac philosophica*), written under the pen-name of Lothario Sarsi, ostensibly one of his pupils (6: 107–9). He immediately brought a copy to Ciampoli on 18 October (12: 494), so he had no intention of hiding his authorship from Galileo, though Galileo found it hard to credit that the work was really Grassi's (12: 498).

Grassi's reply certainly intensified the controversy. Drake, in his very useful treatment of the dispute, even finds that in one passage of *The Balance* deliberate distortion is evident, and suggests that the key to it may lie in Grassi's assertion: 'I wish to say that here my whole desire is nothing less than to champion the conclusions of Aristotle.' Yet I think the passage means: 'I wish everyone to witness that here the least of my wishes is to contend for Aristotle's opinions.' The Latin is certainly tricky, but my translation is supported by the next clause: 'at present I do not delay over whether the sayings of that great man are true or false'; for the moment, he continues, he is only trying to show that Galileo is mistaken on a particular point. If Drake's interpretation is mistaken, it is not thereby a deliberate distortion. Similarly it is perfectly possible to read *The Balance* without finding anything dishonest in its frequent misunderstandings of Galileo or anything more sinister than a general commitment to a modified but still very vulnerable Aristotelianism.[1]

In general the work is quite dignified. Grassi is hurt by Galileo's low opinion of the Roman College, especially after the handsome treatment he was given there in 1611. (From Galileo's private notes on *The Balance* it is clear that he was looking for a fight, claiming that Jesuits wrote whole books against him without a word of praise (6: 115). This is not the calm Galileo of the *Letters on Sunspots*.) Grassi was obviously entitled to defend what he had written, which he did as best he could. He was also free to counter-attack, if he chose. But it would not be easy to pick holes in the *Discourse*, since it committed Galileo to very little beyond the claim that his conjectures and queries merited serious examination, a fact which Grassi may not have understood properly. Any counter-attack would expose him to the full force of Galileo's ferocious polemical skills. Now Copernicanism was not obviously relevant to the debate about comets, but the *Discourse* had called Tycho's system imperfect and said it was important to find the true constitution of the universe (6: 98–9). To that extent it was quite natural for Grassi to refer to Copernicanism, but to needle Galileo on this very delicate topic was perhaps tactless, even with fulsome acknowledgements of his orthodoxy and piety (6: 116, 145). Grassi was mixing theology and astronomy in the very way that Galileo's *Letter to the Grand Duchess* had tried to scotch. Grassi, of course, accepted the ruling of the Congregation of the Index in 1616; not only could he not look favourably on Copernicanism; he would be on the alert for any attempt on Galileo's part to denigrate the rival system of Tycho. His efforts to show that he had understood perfectly well all along how telescopes work was a bit laboured and could be seized on as an attempt to teach Galileo himself (6: 123). Even an innocent and friendly reference to the telescope as Galileo's pupil, if not his child, would incense the prickly Galileo. Grassi wins sympathy when he mentions his efforts, which had seemed successful, to assure Galileo through friends that he had never had any thought of injuring him and he makes a telling point when he says that the upshot showed that Galileo would rather lose a friend than an argument (6: 127). Grassi was no fool. He was a gifted man, a competent teacher by contemporary standards and shrewd enough to realize that Galileo was out to pick a fight (6: 144). Some of his arguments were shaky or so naïve that they were open to devastating mockery, but his views on comets were sensible enough: in following Tycho and most other astronomers he was in fact pursuing what turned out to be the most promising line of research, whereas Galileo's ingenious caution, and habitual aversion to Tycho, led to a dead end on this occasion. Grassi's only misfortune was that if Galileo chose to reply it would be because he was looking for an opportunity to teach the world the proper approach to scientific questions. Guiducci published his own reply in June 1620 and announced Galileo's (6: 181–96): it did not appear until October 1623, after much discussion among the Lynceans from the moment Grassi's book appeared until the finishing touches were being put to Galileo's (6: 199–372).

Galileo was undoubtedly in an awkward position in that he could not argue freely after the decree of 1616. In May 1620 the Congregation at last brought out its detailed 'corrections' of Copernicus's book: they all boiled down to an insistence that Copernicanism be treated as no more than a series of calculating devices (19: 400–1). These corrections did not make things any more difficult for Galileo, but they were a reminder that he would have to be careful. He was not, however, in any sense in disgrace. Sufficient evidence of this is provided by a Latin poem in his praise, entitled *Harmful Adulation* (*Adulatio perniciosa*), sent to him in August 1620 by a long-standing admirer with a reputation as a skilled amateur poet, Cardinal Maffeo Barberini.[2] The death of Galileo's mother that same month was probably as much a relief as a bereavement (19: 443); a year earlier Michelangiolo, surprised to hear that she was as terrible as ever, had offered the bleak consolation that her broken health meant that there would soon be an end to quarrels (12: 494). Sagredo's death in March was a real bereavement, perhaps the greatest Galileo had suffered so far. The death of Cosimo II at the age of thirty in February 1621 was also something of a blow, since it deprived Galileo of a friendly patron; not that Ferdinando II was ill-disposed, but he was only ten years old. Even when he took power in his own name in 1627 his ability to protect Galileo would be very limited.

Galileo's Manifesto

That *The Assayer* is a masterpiece of polemic and a classic of Italian prose is generally agreed; that it was successful by the idiosyncratic standards which govern controversial literature can hardly be denied. Not that Grassi surrendered – there was no reason why he should – but

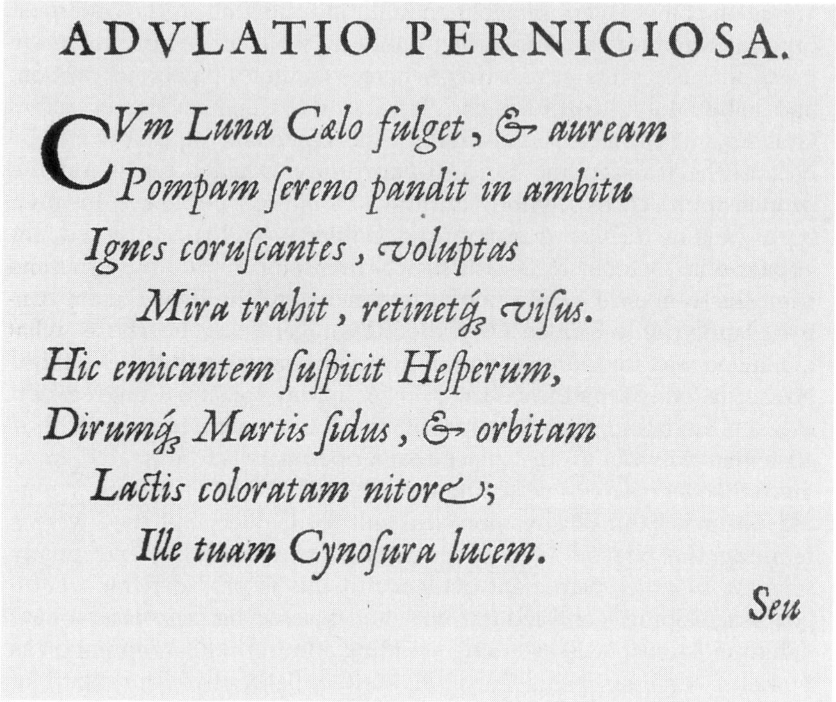

ADVLATIO PERNICIOSA.

CVm Luna Cælo fulget, & auream
Pompam ſereno pandit in ambitu
Ignes coruſcantes, voluptas
Mira trahit, retinetq; viſus.
Hic emicantem ſuſpicit Heſperum,
Dirumq; Martis ſidus, & orbitam
Lactis coloratam nitore;
Ille tuam Cynoſura lucem.

Seu

21 The beginning of Maffeo Barberini's (Urban VIII's) poem in praise of Galileo in the Antwerp edition of 1634.

his lengthy reply of 1626 has been largely disregarded, mainly because its worthy dullness is boring after *The Assayer*. Galileo could safely leave it alone. Grassi is sometimes remembered as the gifted architect of the church of Saint Ignatius in Rome: more often he is the poor Jesuit who took on Galileo. Fairness hardly comes into it, though Galileo was not as unfair as was Pascal in his savage attack on the Jesuits some thirty years later in the *Provincial Letters*; few readers of either classic will have embarked on a study of its fairness. Galileo, it should be said, did have the better of the arguments with Grassi on the ground on which he had chosen to fight. He did not have to show that his opinion of the nature and location of comets was correct or that Grassi's was wrong. He merely had to emphasize the weak points in Grassi's case and show that there were reasonable grounds for saying that we know little about comets. He also had a command of mockery and the maliciously telling image which Grassi (though not Scheiner) might have thought it sinful to employ if he had possessed it. To give himself as free a hand as possible Galileo, quite legitimately and on the advice of his fellow-Lynceans, adopted the convention that he was not replying to Grassi: he was writing a letter to a friend, Virginio Cesarini, about the views of Grassi's young pupil, Sarsi. There had been a suggestion that it should be addressed to Grienberger, but it was decided that it would be too embarrassing for him. The Lynceans seem to have set great store by the appropriate literary form: they wanted a reply to the Jesuits which would demolish their scholastic philosophy but offend the Society as little as possible. It is difficult to take their literary conventions as seriously as they did. Whatever form he chose, what Galileo wanted to say and his characteristic way of saying it would be almost certain to offend Grassi and his fellow-Jesuits. Yet, for all the care he lavished on it, the routing of Grassi was little more than an occasion to expound the true method of science. He was not free to advocate heliocentrism, but he could teach a method which would inevitably transform all natural philosophy and thus, sooner or later, establish the true system of the universe and demolish Aristotelian physics.

Galileo opens with a bitter complaint about the way many have received his publications with hostility: his *Sidereal Message* and the work on floating bodies were attacked fiercely – a rather curious thing to mention, since both had been defended at the Roman College. The *Letters on Sunspots* provoked ridiculous opinions from some – this is presumably meant for Scheiner – and dishonest claims from others, who read his writings and then claimed to have discovered sunspots themselves earlier; this accusation could not fit Scheiner, but most readers would think it was meant to, and the Lynceans certainly thought Scheiner discovered sunspots after Galileo. Capra, Galileo continues, tried to steal the geometrical compass; his teacher, Simon Mayr, who instigated that theft, also published a fraudulent claim to be the discoverer of the Medicean planets. Galileo had resolved to publish no more, but now Sarsi has come up with something new: instead of stealing Galileo's work he fathers someone else's (Guiducci's) on him. So he has to write, though he will leave Sarsi with his mask (6: 213–20).

A little later, in a well-turned insult, he says that he has too much respect for the Jesuits of the Roman College to credit Sarsi's claim that he is speaking for Grassi. But he finds that he has to disagree, however reluctantly, with some things that Grassi himself holds (in the original lecture). The handsome treatment that he received at the Roman College surely does not preclude him from criticizing the views of an individual Jesuit. This perfectly reasonable position is rather spoilt by the additional remark that if Sarsi were a spokesman for the College then Galileo must be held in very low esteem there: worse, if rumour is to be believed, there is open boasting there that they can annihilate everything of Galileo's. Not that Galileo believes this, he adds immediately (6: 225, 227–8), but of course he *did* believe it. Any reader could see that the Lynceans were challenging the teaching of the Jesuits' greatest college and asking people to take sides.

Sarsi's remarks about Copernicanism had to be diverted. The Lyncean Academy was not committed to the new system, but Galileo was and the *Discourse* had mentioned the need for a better system than Ptolemy's or Tycho's. In an adroit move Galileo says that Ptolemy and Copernicus are irrelevant to the debate on comets, since they did not deal with the topic. In any case Sarsi seems to think that in philosophy it is necessary to find a celebrated author to follow. This is shrewd and not unfair, given the Jesuits' official commitment to Aristotelianism. The fact that in this instance Grassi's views on comets were less reminiscent of Aristotle's than were Galileo's could not block the opening Galileo had contrived. Philosophy, he tells Sarsi, is not like the *Iliad* or *Orlando Furioso*, books in which the least important thing is whether what is written is true.

> Philosophy is written in this very great book which always lies open before our eyes (I mean the universe), but one cannot understand it unless one first learns to understand the language and recognize the characters in which it is written. It is written in mathematical language and the characters are triangles, circles and other geometrical figures; without these means it is humanly impossible to understand a word of it; without these there is only clueless scrabbling around in a dark labyrinth (6: 232).

This passage, more than any other, is what makes *The Assayer* a manifesto. If it seems to promise rather more than Galileo had been able to deliver publicly up till then, the pledge was to be redeemed handsomely in the *Discourses*, a book which did not put an end to Aristotelian treatises on physics but did make them obsolete.

Although he had little room for manoeuvre, Galileo moved immediately to undermine Tycho's claim to a following. Ptolemy and Copernicus had at least produced complete systems, Tycho only the unfulfilled promise of one. In any case, not even Copernicus was able to refute Ptolemy: it took Galileo and his telescope to do that and to confirm the superiority of the Copernican hypotheses. Moreover, if the

Church had not come to our aid and shown us that Copernicus is mistaken, we could not have established this from Tycho. So, given that Ptolemy's system and that of Copernicus are certainly false and Tycho's non-existent, why should Sarsi blame Galileo for wishing that the true system should be discovered? This is as cool a coded defence of Copernicanism as anyone could have got away with in Galileo's circumstances (6: 232–3).

Galileo prints the whole of *The Balance* to show how much fairer he is than the selective Sarsi and to demonstrate to the reader how to dismantle an opponent's case. Most of Sarsi's detailed arguments he answers quite satisfactorily by showing that what Sarsi took to be assertions were no more than conjectures attractive enough to be worth consideration. He makes the most of the opportunity to correct misunderstandings about the telescope and ridicules the attempt to explain away the original mistake about magnification being proportional to the distance of the object viewed. At one stage he says Sarsi's argument will bring him no more profit than came to the man who inquired which gate of the city he should go through to get to India the shortest way; in another place he says Sarsi is pecking around like a blind hen (6: 266, 368). In a typical breezy summing-up of one topic he says:

> See the great expenditure of words by Sarsi and me in inquiring whether the concavity of the lunar sphere, which does not exist, by its circular motion, which it has never had, drags with it the element of fire, which we do not know to exist, and with it the exhalations which ignite and set fire to the material of the comet, which we do not know to be in that place and are certain is not stuff that burns (6: 329–30).

He enjoys himself over Sarsi's belief in reports that the Babylonians cooked their eggs by whirling them round in slings: if this does not work for us it must be because we are not Babylonians. Sarsi can believe accounts of cannon-balls melting in their very brief flight, when everyone knows they would last longer in a furnace. Sarsi does not want to insult the authors of such accounts by disbelieving them; Galileo does not want to be so ungrateful to nature and God, who gave him sense and reason, as to value those great gifts less than human fallacies and make the freedom of his understanding the slave of someone who can err like him (6: 340–1).

This soon leads in to a much-quoted passage on what later came to be called primary and secondary qualities. The fundamental idea, which has been enormously influential in philosophy and goes back to the ancients, is that the essential properties of a material thing are shape, size, situation in place and time, motion or rest, contact or lack of contact with another body, and unity or multiplicity; without such qualities bodies are unthinkable. Other qualities are not essential: colour, taste, sound, smell do not belong to the thing itself: they are nothing more than names, except in the animal doing the sensing; they no more belong to the object than tickling resides in a feather. We can be tickled; marble statues cannot. This is not just an attack on Aristotelian verbalism, which was content to refer to a quality of hotness as though it explained something. It introduces what is almost a profession of faith that the true explanation of heat (and many other things) will one day be found to lie in the movements of invisible tiny bodies which make up the objects we can see (6: 340–1). This atomism, in various forms, is an important feature of seventeenth-century science. Galileo did not pretend that he had much in the way of concrete results to make it plausible. He simply asserted his belief that this kind of philosophy would lead to useful discoveries whereas Aristotelian verbalism was futile. That much must surely be granted to him. The fact that the distinction between primary and secondary qualities is itself seriously suspect is something which cannot be pursued here.[3]

A Wonderful Juncture

The Assayer had taken a long time to complete, mainly because of Galileo's recurring illnesses; getting it ready for the press also took a long time, with various Lynceans suggesting improvements. It was very nearly ready when the Lynceans learned to their delight that Maffeo Barberini had succeeded to the papacy on 6 August 1623, after the short papacy of Gregory XV, who had been reckoned to be pro-Jesuit. Naturally there was a flurry of letters and a quick substitution of a new title-page to dedicate the work to Urban VIII. In letter after letter the euphoria of the Lynceans is unmistakable: their time had come. Even before *The Assayer* appeared Galileo asked Cesi's advice about whether he should go to Rome to talk with the Pope at a suitable opportunity. His guarded language is a sign that he was hoping to persuade the Pope to relax the ban on Copernicanism: if that could not be achieved in this wonderful juncture there would be no hope of a similar chance in his lifetime (13: 135).

As soon as the book appeared, even before Galileo received his own copies, it was being read at the Pope's table and pleased him greatly. Two Lynceans, Ciampoli and Cesarini, were well enough placed in the

22   Urban VIII, from *Serie di ritratti d'uomini illustri toscani*, volume 4
(Florence, 1773).

papal court to be able to arrange such things and even to influence
Urban (13: 141, 145). It was harder for Galileo to work out from the
conflicting reports which reached him what Grassi and his Roman
colleagues thought. Grassi certainly began to cultivate Guiducci, who
happened to be in Rome. Guiducci came to nurse hopes of winning the
Jesuit round. One major difficulty was that Grassi could not see how
the everyday motions which we see would be possible if the Earth were
moving. After all, we know that an object dropped from the top of the

mast of a moving ship will not fall at its foot, despite the fact that a fellow-Jesuit had heard from Galileo in his Paduan days that it would. Guiducci, on the way into a church service, assured Grassi that it would and that experiments proved it. Grassi came out of church convinced. That was in October 1624, after ten months of friendly contact: at that stage Guiducci could hope Grassi would not continue the controversy and might even make further progress towards Copernicanism (13: 205–6; 6: 545).

In the meantime Galileo had come to Rome. He arrived on 23 April 1624, after two weeks of consultation with Cesi at Acquasparta. The next day Urban granted him a very friendly audience of an hour, the first of six during Galileo's Roman visit. Despite his high hopes, Galileo was feeling his age. No longer the tireless campaigner of 1611, he lacked the energy to play the courtier to the full (13: 174). The fact that Cardinal Hohenzollern intended to speak to Urban in favour of Copernicanism kept hope alive through May. But Father Riccardi, nicknamed 'The Monster', urged caution: his licence for *The Assayer* had been more like a publisher's blurb than an official clearance, so his partisanship could not be faulted, but he advised that this was no time to revive the Copernican dispute, presumably because Europe was ravaged by war (13: 181). On 8 June Galileo reported his final audience with Urban the day before: the Pope had promised a pension for his son and given him a picture and fine medals. As for Copernicanism, Hohenzollern had put the case strongly and the Pope had replied soothingly that the Church had not condemned it as heretical and would not do so; it was only rash; on the other hand there was no cause to fear that anyone would ever prove it true (13: 179). Urban's friendliness towards Galileo and his admiration for his abilities were further shown in an official testimonial he sent the same day to the Grand Duke (13: 182–4). Where did that leave Copernicanism?

Urban gave Galileo no permission to ignore the decree of 1616 or to bring in theological arguments, yet Galileo felt that he had been given sufficient encouragement to start writing a book about the system of the world. Galileo's understanding, when he left the Pope in 1624, seems to have been that he was free to write a work which would compare the systems of Ptolemy and Copernicus, without taking sides; he would also be free to discuss his theory of the tides. That was certainly what he understood in 1630 when he was negotiating for permission to publish (14: 289). It is also, in form at least, what he did. Urban had no need to confine Galileo to a comparison of calculating devices, since he had no fear that Copernicanism could be proved true. His confidence rested on an argument which he had already tried on

Galileo before he became Pope: no matter what ingenious proofs were brought forward, it was always possible that God had made things otherwise. Since Galileo had not contested this argument from the Cardinal, he was not likely to quibble about it with the Pope, especially since it seemed to leave him sufficient scope for what he wanted to do. Yet this very 'argument of Urban VIII' brought much of his later troubles on him.[4]

I suspect that serious ambiguity was built in to all their exchanges, however friendly. The ambiguity comes in the concept of *hypothesis*: discussion of mere calculating devices was explicitly allowed by the Congregation of the Index; the refutation of objections from Aristotelian physics about motions on Earth would not alarm Urban, since he could always rely on his argument. Galileo, of course, had no time for mere calculating devices, which would reduce Copernicanism to an ingenious fiction. In order to conform to the decree of 1616 he would pretend that he was comparing rival fictions impartially, or at most removing mistaken objections to a realist version of Copernicanism, but in his own mind he would be assessing which hypotheses were more likely to be true and would go as far as he thought safe in dishing the geocentrists, without actually saying that Copernicanism was the true system. He could always cover himself by wheeling out Urban's argument to show that Copernicanism could never be proved.

Letter to Ingoli

On his return to Florence Galileo felt confident enough to write a reply to Francesco Ingoli's attempted refutation of Copernicanism (6: 509–61). Ingoli's manuscript disputation had circulated early in 1616 (5: 403–12). It arose out of oral debates between the two when Galileo was in Rome in 1615; the idea then had been that Galileo would reply to it. But by March 1616 Galileo could do little more than keep up appearances in public. Whatever he made of his interview with Bellarmine, he was certainly chastened enough to realize that he could not flout the intervention of the Congregation of the Index by defending Copernicanism openly. The fact that he felt it safe to write a reply to Ingoli in 1624 is proof that he considered circumstances much more favourable for Copernicanism, but even his habitual optimism did not blind him to the need for reasonable caution. He followed Guiducci's explicit advice to reply only to Ingoli's mathematical and philosophical arguments (13: 186). Again one notices the inevitable ambiguity: even if Galileo kept clear of theology altogether, any suggestion that Copernicanism was more than a series of calculating devices would bring him onto dangerous ground. Since the reply to Ingoli was, in effect, a draft of part of the later *Dialogue* it is interesting to see how Galileo handled this tricky challenge. The manuscript was completed during September and copies circulated widely in Rome, though Ingoli himself was not given one.

Galileo took the ingenious line (repeated in the *Dialogue* with official approval) that the condemnation of Copernicanism did not rely on the sort of arguments used by Ingoli. Protestants should not be misled into thinking that Catholic teaching rested on such fragile supports or that Catholics' attachment to scripture and the Fathers was the product of ignorance of astronomy and natural philosophy. This was ostensibly, even ostentatiously, pious, but it was also a neat way of insinuating that the scientific arguments deployed by churchmen against Copernicus were generally worthless. Catholics, continued Galileo, ramming home his point, need not fear mockery if they place faith in the sacred authors above all the reasons and observations of astronomers and philosophers put together. He expressed his humble gratefulness for the light brought by superior sciences (theology) to the blindness of human reason and wisdom (6: 510–12). This was not mere irony or cynicism on Galileo's part. He did believe that the truths of the Catholic faith could never have been arrived at without divine revelation. His disingenuousness came from having to conceal his conviction that treating the Earth as a planet had nothing to do with faith. His professed loyalty to the official ruling of the Church might be viewed with well-founded suspicion. If he had published the letter at this time he might have got into trouble, or he might have got away with it. But he clearly thought that he had hit on a formula which could give him a lot of elbow-room and he liked it well enough to base the whole strategy of the *Dialogue* on it. Needless to say, anyone who was content to accept the purely 'hypothetical' status of astronomy would not think there was even an apparent conflict between astronomy and theology. Such persons would realize that Galileo wrote the letter only to further the cause of Copernicanism as the true system of the universe. Any lingering doubts left by his soothing introductory pieties would hardly be dispelled by his assertion that it is clearer than the Sun that the Sun, not the Earth, is the centre of the universe, that is the centre of the planetary orbits. A characteristic reservation about whether any place can be called the centre of the universe would have alerted even the most benign reader to the fact that it was not Tycho's compromise system that Galileo had in mind (6: 539). It is remarkable that Galileo thought it safe to fly this kite in a manuscript he allowed to circulate fairly freely.

Galileo easily exposed Ingoli's ignorance of astronomy. More interesting are the passages in which he showed, sometimes almost incidentally, how he viewed Copernicanism at this time. Whether the universe is spherical was, he wrote, unknown: the fixed stars may not be attached to a sphere. He had no difficulty in accepting that the universe is much larger than people had previously thought; it may even be infinite (6: 518, 522, 528–9). He was quite convinced that, as far as human reason went, he had proofs of Copernicanism that no one else had discovered (6: 543). This probably referred to his theory of the tides. But he was clearly not worried if Copernicanism did not fit the facts perfectly. If a mistake or an anomaly was enough to rule out a hypothesis or a system there would never have been any astronomy at all. Corrections and improvements would always be needed (6: 533–4). He was not afraid to stand comparison with Aristotle and his multitudinous followers: a father with twenty sons may not be more fertile than his son who as yet has only one (6: 538).

On the vexed question of what would happen to a stone falling from the mast of a moving ship, Galileo accused Ingoli and Tycho of making up the facts to suit their needs, whereas he had observed for himself what actually happens, though he knew in advance that the stone would fall at the mast's foot whether or not the ship was moving (6: 545–6). Similarly he was sure what would happen to cannon-balls fired in various directions, though here he had no personal observations to offer. It was important for Galileo that Tycho's credibility should be undermined (6: 554). Why believe someone who lied? The lie, of course, was claiming to have observed something that Galileo knew had not been observed. This is a rather crude debating point: by a similar criterion some of Galileo's reported experimental results would look less than truthful. But Tycho, though long dead, was a dangerous rival. He had, according to Galileo, appointed himself umpire and regulator of all astronomical matters. Now there were some good reasons why Galileo should wish to erode the widespread admiration for Tycho: chief among them was that the system outlined by Tycho seemed physically absurd. But some who sided with Tycho were shrewd enough to suspect that Galileo himself was claiming something very like the unique authority he was so anxious to deny to Tycho. Already Galileo was narrowing the debate about the system of the world to two chief rivals, the Ptolemaic and the Copernican. The letter to Ingoli showed much of the shape of the later *Dialogue*. It even included a preliminary draft of one of the most celebrated passages in the *Dialogue* in which Galileo used the example of a windowless cabin in a ship to show that motions inside the cabin would appear the same whether the ship was moving or stationary. Similarly, motions on or near Earth would look the same whether or not the Earth itself is in motion, so most of the favourite objections from Aristotelian physics were simply irrelevant (6: 547–9).

Eucharistic Theology

Although Galileo decided he could ignore Grassi's reply to *The Assayer*, one passage in that book of 1626 does call for comment. Grassi wrote rather archly that he did not propose to examine Galileo's digression about primary and secondary qualities. It seemed to be based on atomism, but it was for the proper authorities to judge its soundness. Still, he ought to air one or two worries. In the Eucharist the substance of bread is converted into the body and blood of Christ: what the senses perceive is merely the accidents such as whiteness. But for Galileo these accidents are just names. Some slippery and cunning minds may think this strange view can be squared with defined Catholic faith by resorting to an arbitrary reinterpretation of authoritative teaching. But if that sort of reinterpretation was not allowed in the case of Copernicanism, still less, Grassi suggested, will it pass when the summit and summary of Catholic faith is concerned. Galileo's annotation on this passage points out that his *Assayer* has already been passed by those whose job it is to judge its soundness: they will have thought of how to remove Grassi's worries (6: 486–7).

Grassi's appeal to Church authority could be seen as a routine implementation of the Jesuits' educational policy: nothing that might conflict with the Catholic faith should be taught or allowed to pass unchallenged in philosophy. At the same time that policy had provided no support to Galileo over a decade earlier when he was sure that the officials of the Church were being hurried into an ill-advised mistake. If Grassi had managed to provoke another authoritative intervention on this new topic, Galileo would have had as little room for manoeuvre as he had had over scripture and Copernicanism – less room, in fact, as Grassi was careful to insinuate. It is no wonder that Galileo put his trust in ecclesiastical princes rather than argument.

He did so with reasonable assurance in this case because Guiducci had told him in April 1625 that *The Assayer* had been examined on behalf of the Inquisition by a sympathetic expert, Giovanni di Guevara: according to Guiducci, Guevara's judgement was entirely favourable

(13: 265). Admittedly both Galileo and Guiducci understood Guevara to have been briefed to examine the book's attitude to Copernicanism. It was only Grassi's book that adverted them to possible difficulties over Eucharistic theology, but they could reasonably tell themselves that a book which had not only delighted the Pope but also passed an Inquisitorial check after publication was pretty secure. In this they were almost certainly correct.

Mention should be made, however, of Pietro Redondi's thesis, based on a document he himself discovered in the archives of the Inquisition. Redondi tries to show that the real, though undisclosed, reason for the condemnation of Galileo in 1633 is to be found in the incompatibility of Galileo's atomism with Catholic Eucharistic doctrine. The strength of Redondi's book lies in his thorough investigation of the theological works of the time, yet his startling thesis has not been accepted. His newly discovered document does indeed show that prior to Grassi's public queries someone wrote in very similar terms to an unknown addressee, presumably an official of the Inquisition. There is no adequate evidence that Grassi himself was the writer and it is not known how it was followed up, if at all, unless we take it that this was the document which gave Guevara his brief for examining *The Assayer*. Redondi's arguments to show that the real reason for Galileo's condemnation in 1632 was the Pope's rage at being made to look soft on Eucharistic unorthodoxy are flimsy. The unsurprising view that Galileo was condemned for his Copernicanism leaves no puzzling gap, certainly not one that would be filled by Redondi's conjecture. Yet, though Redondi has made too much of this particular document, it *is* rather surprising that, as far as we know, Galileo experienced no difficulties over the implications of his philosophy for Eucharistic doctrine. Controversy over this issue began soon afterwards but does not concern his biography.[5]

Vincenzio di Michelangiolo

As head of his family Galileo was frequently put to considerable inconvenience and expense in order to help his improvident brother Michelangiolo. In 1627 Michelangiolo was hoping to leave Munich and settle with his family back in Florence. As a temporary measure it was agreed that his wife, Chiara, should come, with perhaps one child, to be Galileo's housekeeper. Galileo seems to have got more than he bargained for, since most of the children were left with him and the domestic arrangements did not work out well. Michelangiolo's self-pitying letters are the main source for this unhappy episode, which ended in 1628 with Michelangiolo taking his family back to Germany. One aspect of this attempt to help Michelangiolo was not just inconvenient. It was both thankless and potentially dangerous. Galileo transferred the ecclesiastical pension he had obtained from the Pope for his son Vincenzio to Michelangiolo's son, also called Vincenzio. The pension was to support him and pay for musical studies in Rome, since the twenty-year-old Vincenzio's natural ability seemed to point him to the profession of his father and grandfather. The faithful Castelli was now resident in Rome, called there by Urban VIII, who recognized his singular abilities as a mathematician and civil engineer. As proxy uncle, Castelli found lodgings and a music teacher for Vincenzio. But it turned out that Vincenzio was not only idle, unteachable and insolent; he was also ostentatiously irreligious. It was not just that he resented the clerical dress and the set prayers that went with his pension, nor that he detested sermons and put off his annual confession until the last possible moment, nor even that he stayed out all night in dubious company. It was his contempt for all religion that shocked Castelli. Vincenzio could not see why he should join others in worshipping a piece of painted wall. Vincenzio's landlord explained that, if Vincenzio meant those words seriously, the next step was denunciation to the Holy Office: Vincenzio would be burned alive in the Campo di Fiori – shades of Giordano Bruno, whom Galileo never mentioned. Reporting this episode, Castelli urged Galileo to remove Vincenzio to Florence and, if necessary, denounce him to the ecclesiastical authorities there. Even if Vincenzio was bent on his own damnation, at least Castelli could make sure that no harm came to Galileo by association. The only thing to do was to remove the scapegrace from Rome, indeed from Italy, as soon as possible, even if it meant forgoing the money from the pension.

Castelli was deeply shaken by Vincenzio's behaviour, but his forthright letters to Galileo ensured that a public scandal was averted. Galileo must have reflected that he had had little but trouble from his brother's family. (In fact, they returned to Munich. Michelangiolo died at the beginning of 1631 (14: 209).) In this particularly troublesome episode, marked as usual for Galileo by recurring illness, Castelli offered a novel relief: tobacco. His first mention of it must have prompted Galileo to request further information, because Castelli endorsed its therapeutic qualities with a neophyte's zeal. As far as we know, Galileo struggled on without its help (13: 421, 427–31, 435, 443–4).

The brightest aspect of his family life was the affection of his daughter, Sister Maria Celeste. Her letters to him begin about the time he completed the *Assayer*. They are full of solicitude for his chronically precarious health; they are often accompanied by little presents of home-made cakes, or clothes she had mended for him. Occasionally she makes requests for herself or her convent, requests which reveal both how poorly they lived and how little she looked for to keep her happy and supportive of others, especially her depressed sister. Now and again she proffers gentle spiritual advice, secure in the knowledge that it will not be taken amiss. Galileo's later move from Bellosguardo to Arcetri was her idea, first mentioned in May 1630, but we can be sure it was as much for his sake as hers that he went there.

While he was completing his *Dialogue* Galileo heard the good news that Cavalieri had been appointed professor at Bologna in August 1629. Castelli's pupil gratefully acknowledged Galileo as his teacher. Galileo, appreciating that Cavalieri was taking mathematics into new regions, ones that he himself could not hope to explore, did his best to persuade the Bolognese authorities not to swamp Cavalieri with chores like compiling astronomical tables (14: 43).

The Dialogue

Galileo was able to write to Cesi for Christmas 1629 with the news that he had completed the body of the *Dialogue*, despite the fact that his sight was failing (14: 60). A couple of weeks later he was planning to go to Rome to arrange publication by the Academy (14: 67). In February Castelli reported that Cardinal Francesco Barberini was uneasy about the argument from the tides, since that would make the Earth a star (14: 78); in March he sent the more encouraging news that if it had been up to Urban the ruling of 1616 would never have been made (14: 87–8). Anyway the person responsible for licensing books was Riccardi, who had puffed *The Assayer* heartily in 1623, so the prospects were good. Galileo visited Rome from 3 May to 30 June. Despite a friendly audience with Urban, he discovered that the Pope was unhappy about the argument from the tides; though he did not demand that it should be dropped, he did insist that the book should not be called *On the Ebb and Flow of the Sea* (14: 105, 113). Still, Galileo left Rome confident that there was no obstacle to going ahead with publication, except that Cesi was ill (14: 130). News that Cesi had died unexpectedly on 1 August was a severe blow. What had looked like being a fairly brief and straightforward negotiation for permission to print turned into a protracted series of moves and counter-moves, greatly complicated by the outbreak of a dreadful plague which interfered with correspondence. Galileo, with the help of pressure from the Tuscan ambassador in Rome, was able to get permission to have the book censored and published in Florence, provided it conformed to instructions sent by Riccardi. He presumed, perhaps innocently, that this entitled him to claim that he had Riccardi's permission to print, so when the book appeared in February 1632 it included Riccardi's *imprimatur* along with the Florentine Inquisitor's. That would be a cause for complaint, since Riccardi's *imprimatur* could apply only to books printed in Rome. Had Galileo gone too far in other ways as well?

He clearly thought he had the goodwill of the Pope, provided that he acted in concert with Riccardi and carried out his instructions. More than once during the negotiations he pointed out that he was prepared to label as dreams, chimeras, equivocations, paralogisms and nullities any reasons or arguments which the authorities thought too favourable to dubious opinions (14: 259); such labels are indeed scattered liberally throughout the *Dialogue*. Now Riccardi seems to have been satisfied by this. He insisted, on the Pope's orders, that the tides should not figure in the title; he also required that the book should begin and end with suitable disclaimers (which Galileo included). But he described the work to the Florentine Inquisitor as discussing Copernicanism 'probably'. It is not entirely clear what he meant by this, nor whether he endorsed such a procedure; he certainly did not object to it. In any case, he passed on the Pope's stipulations: Galileo should discuss Copernicanism purely mathematically with a view to proving that, divine revelation and theology apart, the appearances can be saved in this position, solving all the objections from experience and Aristotelian philosophy, but without conceding absolute truth to this opinion (19: 327). One can see why Galileo thought he was not going beyond what would be acceptable to the Pope.[6] One can also see that only some version of instrumentalism stood between the Pope and a two-truths theory, where something could be true in philosophy and false in theology. Galileo, then, had reasonable grounds for thinking that publication was not as risky as it might seem to anyone who did not know he was complying with official guidance. It was essential, of course, not to concede absolute truth to Copernicanism. We shall see in the next chapter how he handled that difficulty.

8

The Dialogue and Galileo's Condemnation

Many readers, from Descartes onwards (14: 124–5), have found the *Dialogue* shapeless and undisciplined: it hops from one subject to another, sometimes leaving even the participants asking each other how they landed where they are. Yet it was precisely because it allowed for lots of interesting digressions that Galileo thought the dialogue form most appropriate for his purposes (7: 30).[1] The participants are Salviati, who presents Galileo's own views, Sagredo, who is host and chairman, and Simplicio, an unreconstructed Aristotelian called after the sixth-century commentator on Aristotle. Since Simplicio's principal function is to rehearse all the difficulties which prevented people from understanding, let alone accepting, Copernicanism, he is not very bright or well-informed; occasionally, however, his preternatural obtuseness makes him little more than a straw man, which is why many have thought his name is meant to suggest a simpleton.

Giovanfrancesco Sagredo, as host in his Venetian palace, is the chairman. Ostensibly open-minded and unprejudiced, he is hardly impartial: he colludes with Salviati in patronizing Simplicio, who is like a local lecturer hauled in to serve as a foil for the distinguished visitor, the noble Florentine Filippo Salviati. Galileo recalls with fondness dis-

THE *DIALOGUE* AND GALILEO'S CONDEMNATION

cussions he had in Sagredo's palace with Sagredo, Salviati and an unnamed Peripatetic philosopher. (This may be a literary device. In 1612 Sagredo seems to have known Salviati only from the reports of others (11: 315). Thereafter there was no occasion when the three could have met in Venice with an Aristotelian.) He also recalls Salviati's hospitality at Le Selve, his villa near Florence, especially when Galileo was writing the *Letters on Sunspots*. Salviati, a fellow-Lyncean, was greatly admired by Galileo, to whom he would seem an obvious choice to be his spokesman, though we have little independent evidence that he had any outstanding scientific ability. Salviati's interventions are supplemented by frequent, usually adulatory, references to Galileo himself as 'our common friend' or 'our Academician', a subtler means of introducing self-praise than the face-to-face flattery of the early dialogue on motion. The book was Galileo's tribute to two close friends whose premature deaths had left him still feeling bereft of their good sense, cheerful company, reliable support and, in Sagredo's case, biting wit. That the *Dialogue* was no dispassionate weighing of current disputes but a partisan marshalling of brilliant arguments would have seemed the most natural thing in the world to them. Whether it should be categorized as propaganda or rhetoric is a question I shall come to later. Whether it was even formally compatible with the ruling of the Congregation of the Index is another important question which I postpone for the moment.

A clue to the underlying shape of the work is found in Galileo's long-standing desire to write a work on the tides: but for the Pope's insistence, it will be recalled, it would have been entitled *Dialogue on the Ebb and Flow of the Sea*. The fourth and final day is devoted to this topic, which for thirty-five years Galileo had seen as a way of proving that the Earth is a rotating planet: indeed, for him it was the principal topic of the book (14: 289). But first he set himself two other important tasks: to discuss whether observation of motions on or near Earth can show that the Earth is fixed, a topic to which he devoted the second day, when the Earth's rotation on its axis is discussed; then to examine celestial phenomena in order to see whether the Earth's revolution round the Sun provides a better hypothetical account of planetary motions than did Ptolemy's geostatic astronomy, the topic of the third day. This is all set out with admirable clarity in the preface, but confusion arises because Galileo does not announce that the first day will be devoted to discussing whether Aristotelians are right to rule out Copernicanism as impossible on the grounds that the divorce between the Earth (with the elements) and the heavens is an essential feature of the universe.

GALILÆI GALILÆI LYNCEI
Dialogi,tam eos quos edidit
DE SYSTEMATE MUNDI
quam quos
DE MOTU LOCALI.

Ptolomæus

Aristoteles   *N. Copernicus*

LUGD.BATAV. *Apud* { FREDERICUM HAARING et~ }
{ DAVIDEM SEVERINUM } Bibliopolas
{ M.DCC. }

J. Mulder Fecit.

23   The frontispiece of the Latin version of the *Dialogue* and *Discourses*
(Leyden, 1700) showing Aristotle, Ptolemy and Copernicus

First Day

The contrast between the unchangeable heavens and the elementary
world was based on the distinction between the rectilinear motion

upwards or downwards proper to the elements and the circular motion proper to heavenly bodies. Galileo is quite explicit that this is the foundation stone of Aristotelian physics (7: 42): he chips away at it from the start of the *Dialogue*, though it is mainly the second day which shows that there could be an alternative foundation for physics.

There are several reasons why the contrast between Earth and heaven provided a good place to start his *Dialogue*. Aristotelians had long been under attack not only from Copernicans but also from geostatic astronomers like Tycho, who saw that the supernova of 1572 was a signal instance of change in the heavens. The comets observed by Tycho, the supernova of 1604 and the comets of 1618 all added to the embarrassment of those who wished to exclude change from the heavens: on this point they were under siege or, in Sagredo's image, they were shoring up their tottering building and securing it with chains (7: 81). Best of all, from Galileo's point of view, was the fact that his own telescopic discoveries provided the most striking, most easily intelligible and most interesting evidence that heavenly bodies are nothing like as different from Earth as had been thought. It was still difficult for people to accept that a stone could drop to the foot of a tower if the Earth is moving. After a couple of decades in which to get used to the ideas, it was at least easier to allow that there are mountains on the Moon and spots on the Sun, whatever Aristotle had said. Galileo was fully conscious not only that the telescope had given him information that was not available to the ancients but that this very fact lessened the authority of Aristotle in matters of natural philosophy, especially motion. A favourite theme of the *Dialogue* and other works is that if Aristotle could come back and see what the telescope shows he would agree with Galileo. Consequently, Galileo was not just scoring a debating point when he said that fundamentally he was more faithful to Aristotle than were his followers, because Aristotle, like Galileo, built his philosophy on sense experience and reason, whereas they rummaged through Aristotle's texts instead of trying to understand nature itself (7: 75, 80–1). Galileo also continued his policy of staking a claim to enemy territory. Celestial physics would no longer be an Aristotelian preserve, with astronomers confined to the computation of motions. To offer an estimate of the height of lunar mountains and to give reasons why there seems to be no water on the Moon was to mark a new chapter in physics and to subvert not only the traditional departmental division of academic labour, but also the world-picture it sustained.

In several passages Galileo is happy to call himself a Platonist (or even a Pythagorean) rather than an Aristotelian, but usually all he means is that without mathematics we shall never understand the real world. He certainly holds that geometry reveals the intelligible patterns which underlie and explain the everyday world of common sense, but this is a long way from Plato's real world of forms or patterns, so different from our shadowy and derivative world of common experience. Nor has Galileo any time for the kind of number mysticism associated with Pythagoreanism (7: 35). He does like to use the Platonic notion that all our knowledge is a sort of remembering. This is an educational theory; in practice it means that even people as thick as Simplicio can understand novel accounts of falling bodies on a mobile Earth if only they will use their native intelligence and not behave as though knowledge is a matter of ferreting out what Aristotle wrote (7: 217). In contemporary terms, of course, Galileo was very much a Platonist: he was one of those who were convinced that mathematics could not only not be marginalized, as it was by Aristotle and his followers, but that it was the key to understanding the world. Yet it would still be reasonably accurate to call him an Aristotelian in aims; what he was trying to do was reform the methods of natural philosophy by bringing together reason and sensory experience in the understanding of what we observe. Simplicio reminds Salviati more than once that Aristotle is the champion of experience, which always has precedence over human reasoning; this (with a notable exception, to be discussed later) is grist to Galileo's mill. To Aristotelians Copernicans were purveyors of paradox who denied plain facts and even the value of the senses themselves (7: 57–8, 80). Galileo knew well that Aristotle insisted that all our knowledge starts from the senses; in physics, however, Aristotle had not had the advantage of later observations. In any case he had been too easygoing in taking unexamined everyday observations as an adequate basis for science, besides being programmatically misguided in confining mathematics to an abstract study not related to the world our senses reveal.

Galileo showed that it was no longer plausible to treat the heavens as completely different from Earth, though it was no part of his programme to suggest that they are exactly like Earth. Sagredo rules out lunar plants, animals and humans, though there could still be other lunar beings subject to change. Salviati does not exclude the possibility that the moon contains beings very different from us who praise the Creator (7: 86–7). In any case, the unchangeableness of the heavens was clearly exploded. So the discussion of the first day left the Aristotelian account of rectilinear and circular motions looking very shaky: would the well-known difficulties which a moving Earth brought with it in physics and astronomy be sufficient to rule out the hypothesis of the Earth's motions?

The Second Day

Galileo could easily show that it is simpler to make the Earth rotate eastwards on its axis than to have all the stars going from east to west each day on a vast sphere, dragging the planets with them. As Sagredo says, with a courteous gesture to the Florentine Salviati, if someone climbed up to Brunelleschi's cupola to see Florence it would be unreasonable for him to want the city to be rotated before his eyes to save turning his head (7: 141). Aristotelians, of course, knew a rotating Earth would be simpler; they just thought it impossible. Simplicio puts their position neatly: the important thing is to be able to move the Earth without a thousand inconveniences (7: 148). Some inconveniences had been pointed out from antiquity; others had been added more recently; all were still being rehearsed in a list which by this time was almost canonical. A stone can be dropped from a tower and land at its foot, whereas if the Earth were moving it would fall far to the west, since while it was in the air the tower would have moved to the east; this argument was reinforced by the claim that a stone dropped from the crow's nest of a moving ship does not fall at the foot of the mast. Similarly arrows (or, in more up-to-date versions influenced by Tycho, cannon-balls) would overshoot, undershoot or be to the side of the target, according to the direction of aim. A rotating Earth would produce a constant gale vastly stronger than what we feel when horse-riding and birds could never fly against it. People, animals, even buildings would be flung off an Earth which, at its equator, would be rotating at 1,000 miles an hour. What Simplicio called 'inconveniences' were clearly impossibilities; a good, empiricist Aristotelian could see that the consequences of a moving Earth did not obtain.

Galileo goes through these difficulties one by one at considerable length, pointing out, for instance, that he knows in advance that Aristotelians are mistaken about what actually happens when a stone falls from the mast of a moving ship. From our point of view, and indeed from his, they were mostly the same difficulty, but he knew that his principal task was to convince Aristotelians that there could be another way of looking at motion. We have here another instance of what is often referred to as a 'paradigm switch'. It is frequently suggested that Aristotelian and Galilean physics are so different that they are not really comparable with each other; they are *incommensurable*; to switch from one to another is a more a matter of sudden conversion than a purely rational process.[2] This is not a suggestion that Galileo himself would have supported, even though he occasionally abandoned hope of convincing particular opponents, as we saw in the dispute over floating bodies. He makes great play with the fact that all Copernicans started out as Aristotelians and were convinced by good reasons that the Aristotelian difficulties could be met (7: 154, 159). We could say that Galileo, and other Copernicans, were bilingual: they understood the language of both the old physics and the new. This does not prove, of course, that there could be a one-to-one translation between the two languages. In the new language some things are no longer thought worth saying or even bothering about. But what Galileo could do was use arguments and images to let Aristotelians see that what they thought were insuperable difficulties were nothing of the kind: there could be another way of looking at stones falling from towers which is just as close to what is observed and yet fits in with a moving Earth. If the Aristotelian allowed this much he had not thereby conceded that the Earth is moving; but he had admitted that the question would have to be settled otherwise than by appealing to Aristotelian accounts of motion on Earth.

Galileo was a brilliant teacher, a devastating controversialist and an entertaining debater, but his defusing of the objections from Aristotelian physics is a rational procedure. No doubt all sorts of factors induced people to adopt Copernicanism; hunches, imaginative flair, ambition and temperament were significant ingredients for many, as they certainly were for Galileo. Nor could Galileo ever produce what could be counted as a conclusive proof that Copernicanism is the true system of the universe. But talk of incommensurability, paradigm switches, or of conversion to the new science of motion and the new system of astronomy should not obscure the fact that rational discussion of the alternatives played an important role. (It is not unknown for it to play a role in religious conversion, which itself can come at the end of a long process.) The second day of the *Dialogue* provides a classic case of such rational discussion of a topic which was crucial to the development of both astronomy and physics. It is sometimes dismissed as propaganda. The word 'propaganda' comes from the Latin name of the Roman Congregation for Propagating the Faith, which was set up in its definitive form in 1622 (with Ingoli as secretary). It is quite accurate to see the *Dialogue* as an effort to propagate or spread the Copernican faith, provided one does not import the modern meaning of 'propaganda' with its implication of dishonesty. It is simply a work of persuasion which uses rhetoric to convey what Galileo thought was true. The *Dialogue* has its shortcomings and questionable simplifications but, generally speaking, it is as honest as it well could be, given the limitations imposed by the Church's intervention of 1616.

One could say, of course, that there was little that was really new in the second day of the dialogue. Copernicus, and even Bishop Nicole Oresme in the fourteenth century, were quite at ease with the idea that we can ignore the motion shared by a rotating Earth and everything on or near it. Galileo could easily have shown how far he had advanced from their ideas, but he chose to reserve most of his own contributions for a later work and was content to stay surprisingly close to Aristotle in his talk of natural and violent motions. Nevertheless, some of his most prized discoveries did get something of an airing in the *Dialogue*. Even a Simplicio could not fail to notice how Salviati could display a range of fruitful and startling ideas merely by fixing a ball on a string, hanging the string from the ceiling and letting the pendulum swing back and forth. When the ball or bob was set in motion by pulling it away from the perpendicular its swing took it to an equal height on the other side of the perpendicular – well, more or less, because things like air resistance had to be allowed for (7: 43, 253, 262). Light bobs, such as cotton balls, come to rest before heavier ones: this is a contribution from Simplicio, who is getting involved (7: 177). The shorter the pendulum the greater the frequency of the swings, says Salviati, and the amplitude of the swing of any given pendulum does not affect the frequency of its oscillations: the period is almost the same whether the pendulum is released from two degrees or eighty – a real poser for Aristotelians (7: 256, 475–6). Galileo could easily have exploited his discoveries about pendulums (and inclined planes) to strengthen what he said about motion in the *Dialogue*, but it made good sense to reserve his own main treatment for the *Discourses* and to concentrate in the *Dialogue* on removing Aristotelian objections. That is why even the law of falling bodies is mentioned only to claim priority for Galileo and to advertise its proper publication in his forthcoming work on motion (7: 248). Like Galileo, I shall postpone comments on his account of pendulums to the next chapter.

Perhaps a great scientist would have been more usefully employed in writing for his peers, but non-scientists will always have a soft spot for those who can explain important scientific issues to them. Moreover, Galileo's peers were by no means all Copernicans: most contemporary scientists (if we may use an anachronistic term), especially those who were Aristotelian philosophers, certainly stood in need of tuition from Galileo, as Grassi had needed help from Guiducci. This elementary instruction, packed into the second day of the *Dialogue*, was an essential prerequisite for an understanding of the more advanced treatment of motion in the *Discourses*.

As he works through the list of traditional objections Galileo tries various ways of convincing the Aristotelians that their difficulties all arise from taking for granted what is in dispute, namely that the Earth is fixed (7: 170–97). Once this inveterate assumption is uncovered it is possible to see things differently. A hand lets a stone fall from a tower and the stone falls to the tower's foot: but the tower, hand and stone all share the motion of the Earth and so does the observer. So the observer is not aware of this shared motion and sees only the vertical fall. One has, of course, to accept what seemed unthinkable to the Aristotelian, namely, that a body can have at once a vertical motion downwards and another (imperceptible) motion imparted to it by the rotating Earth, and that these two motions are compounded with each other without mutual interference. But this seeming paradox was made more easily assimilable by examples. An artist whiles away the time on a voyage from Venice to Syria by drawing a picture: when we look at the picture we do not see tiny deviations from the smooth line which the pen must have traced across the Adriatic. We just see the lines of the picture, which could just as well have been drawn ashore. Yet the pen did trace out a line as the ship carried it over the sea, just as a tower moves with the rest of the Earth, even though in the nature of the case we cannot see this motion. In other words, things look exactly the same whether or not the Earth is moving (7: 198). This simple point is made page after page as each difficulty is worked through. They are all brought together in the brilliant image, sketched out in the *Letter to Ingoli*, which Galileo now elaborates to sum up what he is saying in this part of the *Dialogue*.

Shut yourself in with a friend in a large, windowless cabin or stateroom of a motionless ship, says Salviati. Have flies and butterflies flying round in the cabin, fish swimming in a bowl and a tap dripping water into a narrow vase. Observe all these motions while the ship is motionless; you will see the flies flying equally in all directions and the fish swimming wherever they want in the bowl. You can throw things to your companion with equal ease in any direction. None of this would surprise the Aristotelian, but, says Salviati, now have the ship move as fast as you like but perfectly smoothly. You will not notice any difference; there will be nothing to tell you whether the ship is moving or not. Jumps you make will be the same as before, even though the deck moves in the opposite direction while you are in the air. A throw to your companion still takes the same force whether to prow or stern. No drop will miss the narrow vase and splash towards the stern. The flies and fish will move as easily in all directions as formerly. Lighted grains of incense will make a little cloud of smoke, which is no more inclined to go one way than another. The reason that everything behaves as it did when the ship was motionless is that the ship's motion is common to everything in the cabin. In the same way, all the hackneyed experiences which are intended to prove that the Earth is fixed or that the Earth is moving are useless: they are all compatible with either a fixed Earth or a moving one, just as all the movements in the enclosed cabin are identical whether or not the ship is moving (7: 212–14).

This exposition gives the name to Galilean relativity, though since his ship is on a spherical surface the attribution is partly honorary. Anyway Sagredo agrees that it knocks out all the objections except the one about objects being flung off by the Earth's rotation. On this Salviati is very ingenious but less convincing. He satisfies Sagredo by geometry that a point on the equator may be moving very fast, but this is not what counts, because its angular velocity is small and the Earth's radius is very large, so objects are in no danger of being flung off (7: 223–9). In general, the *Dialogue* strikes a modern reader as replete with brilliant simplifications. Some of these are accounted for by its restricted purpose, which was to clear the ground for Copernicanism rather than to establish a new science of motion. Some, like the assumption that no motion on or near Earth could show whether or not the Earth is in motion, actually conceded too much. Galileo himself realized that a stone dropped from the mast of a moving ship should theoretically fall slightly to the *east* of the foot, since the top of the mast is moving faster than the foot, but he thought the divergence undetectable. Some may come from Galileo's self-confessed joy in discovery and his delight in showing himself more acute than others (7: 237). Others stem from his concentration on pure description of motions without searching for causes or forces to explain them: this concentration served him well, but it left essential physical concepts such as *force* to be worked out by others. If, despite the limitations of Galileo's aims, his new science of motion in the *Discourses* strikes us as recognizably like later physics (in contrast to the work, say, of Borro or Buonamici), then the first two days of the *Dialogue* can easily be seen as a necessary introduction to his detailed geometrical kinematics in the *Discourses*. It need not be regretted as a diversion forced upon him by the exigencies of establishing Copernicanism.

The Third Day

This gets off to a dull start with a lengthy attack on an Aristotelian's attempt to show that the new star of 1572 was sublunar. The digression enables Galileo to explain about parallax and to teach Simplicio, who is

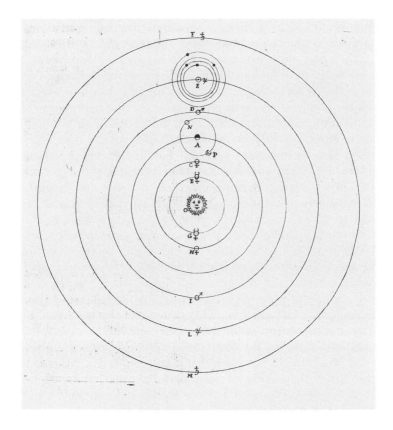

24   A diagram of the heliocentric system from Thomas Salusbury's translation of Galileo's *Dialogue* in *Mathematical Collections* (London, 1661).

taken to be totally ignorant of elementary geometry, an important lesson about how to correct reported observations of stellar positions, but the conversation comes to life again only when they turn to the annual motion of the Earth (7: 312–13). What Galileo offers here is little more than a beautifully simplified exposition of Copernicus's system, reinforced by telescopic novelties, all of which he claims to have discovered first. There is no suggestion that Kepler had transformed the Copernican system. Clearly Galileo was not bothered about closeness of fit. He was so impressed by the Copernican explanation of the retrograde motion of the planets that he thought it should be sufficient to convince everyone. That explanation is, indeed, elegantly simple in outline. One can see how a Galileo or a Kepler would be so impressed that they would think Copernicanism must be true. Yet it was the Pythagorean Kepler, with obsessions which would be dismissed as dotty if they had not spurred him on to his great achievements, who toiled for years until he had made Copernicanism truly Sun-centred and succeeded in replacing the basic principles of planetary astronomy. He took the greatest achievements of 2,000 years of astronomy, based on combinations of uniform circular motions, and with non-uniform motions in elliptical orbits he reformed the system of Copernicus into something simpler and more accurate than any previous astronomer had achieved. Galileo, whose accounts of terrestrial motions could be similarly breath-taking in their elegance, novelty and importance, seems comparatively slapdash in theoretical astronomy. It is as though he could see that Copernicanism must be the true system, so if there were loose ends they could be tidied up later by someone else. This was a perfectly sensible stance: science would hardly be possible if every new idea had to settle all relevant difficulties. But it did make him vulnerable, given that he was actually trying to win the world round to Copernicanism and was supposed to be comparing the strengths and weaknesses of the two chief world-systems. Not that an exposition of Keplerian heliocentrism would have been enough to save his book from condemnation, but it would have impressed posterity more.

It must be admitted, however, that if one could, like Galileo, disregard the clutter, then Copernicanism does have an underlying simple outline which is very striking. Since traditional astronomy plotted planetary paths from a moving Earth which was thought to be fixed, one could look at Ptolemaic astronomy and see in it reflections of heliocentrism. The paths of the outer planets, for instance, must have the orbit of the Earth somehow superimposed on them. In fact the main epicycles of Mars, Jupiter and Saturn, which produce the curious loops of retrograde motion are, to a Copernican, merely aping Earth's orbit. As the Earth overtakes each outer planet there is a sort of optical illusion. It is as though the Earth and the planets were on circular racetracks. From Earth we can plot the path of an outer planet as it travels eastward ahead of us on its own track; as we catch up with it, it seems to slow down and when we overtake it even seems to go backwards (westward) for a while until it resumes it eastward motion in our wake. This, as Salviati makes clear, is a simple consequence of the Copernican arrangement of the planets, with planets moving faster the nearer they are to the Sun. In the case of the inner planets, Venus and Mercury, their epicycles have the role of the planetary orbit, with the deferent playing the role of Earth's orbit: retrogression occurs when they overtake us. The striking thing is that the basic Copernican scheme is itself enough to provide a qualitative explanation of planetary retrogression. In fact Copernicanism is truly a system in which everything hangs together, whereas Ptolemaic astronomy is not a true system but just a very sophisticated method for making predictions about planetary positions. It is not that Copernicus's system was more accurate: it was not, and in any case, in this respect the two approaches (and Tycho's) were intertranslatable, so that one could produce whatever the other could. But because it was a system Copernicanism made sense of some things which to geocentric astronomers were merely coincidental or even contrived, and in any case not worth pursuing.

Every astronomer knew that Venus and Mercury cannot be seen very far from the Sun: this fact was catered for perfectly adequately in Ptolemaic astronomy by tying the centres of their epicycles to the line joining Earth and Sun. That did the trick. But to minds like Copernicus's or Galileo's it was much more satisfying to see that if the Sun is placed in the centre of the universe, with Mercury, Venus and Earth in orbit around it, then no more is needed. The system itself ensures that we can never see those planets far from the Sun. Similarly, Ptolemaic astronomers knew that Mars, Jupiter and Saturn, as they travelled in their epicycles, kept an eye, as it were, on the Sun: a line joining the planet to the centre of the epicycle stayed parallel to the Earth-Sun line. This was just a brute fact for Ptolemaic astronomers, but for Copernicans like Galileo it was inescapable evidence that the Earth revolved round the Sun.

Galileo again makes the most of his telescopic discoveries. The phases of Venus showed that Ptolemy could not be right about Venus: it must be in orbit round the Sun and the same must go for Mercury. (This major modification, as we saw in chapter 4, was not too difficult for conservative thinkers to accept. Indeed, they were by this time quite happy to have all the planets go round the Sun, provided the Sun itself remained in orbit round a fixed central Earth. In other words, Tycho's compromise system was widely acceptable.) But Galileo improves the occasion by praising Copernicus for staying committed to his system even when his senses contradicted his reason. His reason told him that Venus should show phases, yet it did not. Similarly his system demanded very large changes in the apparent sizes of Venus and Mars according to whether they were near or far from Earth. This was not the sort of thing that bothered Ptolemy, because he was a calculating astronomer, but it meant that a philosophical astronomer like Copernicus again had to rely on his reason to believe what his senses contradicted

(7: 367). His vindication came, of course, when Galileo used his telescope to show that things were actually just as Copernicus's reason had determined. So the ultimate test is not just any old sensory experience, because reason was vindicated by a higher sense, not just vision but vision enhanced by Galileo's telescope.

This brilliant piece of rhetoric indicates some of the difficulty in deciding when an anomaly should be classed as falsifying a theory. If one insists that none of this amounted to proof of Copernicanism, that is the sober truth. But it is important to feel some of Galileo's own excitement and to understand why he was so convinced that the true system of the universe had been found.

He also uses the discovery of sunspots to construct yet another strong argument in favour of Copernicanism, one that depends on the fact that the Sun rotates on its axis. It was, it seems certain, only while writing the *Dialogue* that Galileo realized that the axis of the Sun's rotation is inclined to the ecliptic. He tries to show that all his observations of sunspots, with their seasonal variations, then fit perfectly if their paths are viewed from Earth rotating in its orbit. Could there not be a geostatic account? Of course, but it would be so complicated as to be implausible. Put like this it is quite a strong argument, though it seems to have been thrown together rather hurriedly, since as presented it does not fit the facts at all well. Scheiner, as it happens, had specialized in the observation of sunspots. His *Rosa Ursina* of 1630 described the inclination of the Sun's axis and accounted for the observed paths of sunspots much more accurately, despite the complications attendant on a geocentric account. This is not to say that Galileo stole the idea from Scheiner's book: if he was going to plagiarize, why mess up the observational side? Drake is sure that Galileo could not have introduced a major topic at a very late stage in the writing of the *Dialogue*, when he was writing within the limits allowed by Riccardi. Galileo could have got the idea of the inclination of the Sun's axis from a letter written long before by Francesco Sizzi. This may be so: Galileo certainly had plenty of advance warning that Scheiner was going to publish on sunspots, so he may have gone over his own writings and correspondence and realized that he could construct a new argument in favour of heliocentrism.[3]

But in that case one could hardly expect Scheiner to be convinced by Galileo's account in the *Dialogue* of how he made the discovery. Salviati describes how he was with Galileo in his villa 'some years' after the publication of the *Letters on Sunspots*. Their attention was taken by a large, solitary spot, so they tracked it carefully. One day 'our Academic' exclaimed:

Filippo, I think the way to something of great importance has just opened up to us. For if the sun's axis of rotation is not perpendicular to the plane of the ecliptic but inclined to it, as the curved path we have just observed suggests to me, we have such a conjecture about the states of the sun and the earth as has never before been given to us with such firmness and conclusiveness by any other circumstance (7: 374).

This introduction to what Galileo thought was a very effective argument would neatly cut out any claim to priority by Scheiner, preserving Galileo's perception of himself (made two pages earlier) as the sole discoverer of all the heavenly novelties, including sunspots. It is, however, not even a coherent account. Galileo never met Salviati 'some years' after publishing the *Letters* of 1613, for Salviati left Florence that year and was dead the next. One may put this down to literary licence. It does not prove plagiarism, but it leaves a suspicion of incomplete honesty which is not dispelled by Drake's conjecture.

Scheiner and Galileo were in any case irreconcilable before the *Dialogue* was published. Scheiner's massive work on the Sun started with his apologia, an exhaustive attack on the 'Censor' (Galileo). He lacked Galileo's wit but compares well with him for combative bitterness. He was shrewd enough to tax Galileo with wishing to be acknowledged as the sole discoverer of everything new, as though an angel had revealed everything to him. If Galileo makes fun of those who appeal to authority, why then does he expect us to bow to his *ipse dixit*? Galileo will be laughed at like Aesop's frog, which puffed itself up until it burst. Scheiner's apologia, however ill-humoured, does raise awkward questions for those who see Galileo only as the innocent victim of obscurantism.[4]

The third day concludes with discussions of the strongest astronomical objection to the orbital motion of the Earth, namely that it required a universe largely composed of waste space since the stars would have to be far enough away not to show any parallax. Galileo was confident that annual stellar parallax would eventually be detected; he even made ingenious suggestions as to how to go about detecting it. In passing he was able to dispose of the objection that Copernicanism would make the stars ridiculously large (since they look the same size to us, even if they are as far away as Copernicus thought). He showed that this is an illusion: most of what was taken to be the apparent size of a star was composed of adventitious rays. The telescope shows that what might be called their real apparent size, their true visual diameter, is much less (7: 389). But his fundamental argument was: who is Simplicio, who are we, to say the universe is too large? As for talk of waste space, why should we be the sole care of Providence? Why should things be called useless, just because they were not made for our use? The spaces that were to terrify Pascal were not necessarily infinite for Galileo: to the end of his life he remained undecided about whether the universe is infinite. But he was perfectly equable about this: the universe is whatever size God made it.

The Tides

Galileo had long been convinced that the phenomena of the tides would provide a conclusive argument in favour of Copernicanism. He dismissed traditional attempts to explain them by the influence of the Moon. He was astonished that the great Kepler, with his free and acute mind and his grasp of Copernicanism, could believe in the dominion of the Moon over the water, in occult properties and similar puerilities (7: 486). In fairness to Galileo, one should remember that there was as yet no convincing explanation of the tides. His own theory was, in fact, mistaken but it was at least an attempt to provide an explanation. Better still, from Galileo's point of view, it was the one thing, in the absence of detectable stellar parallax, which could show that the Earth was moving. The simplicity of the Copernican explanation of planetary retrogression and of the paths of sunspots was striking, but it could not force anyone to abandon geocentrism. Galileo's defusing of Aristotelian objections to the Earth's motions depended, as we have seen, on the claim that no motion observable on Earth could settle whether the Earth is moving: all such motions would look the same, whether the Earth was fixed or in motion. One would think that to be consistent Galileo could not build any proof of the Earth's motions on the tides, but he was convinced that with the tides it was very different; they provided him with a fitting climax for his book.

If the Earth was immobile there could be no way of accounting for the tides, whereas given the Earth's two motions the ebb and flow of the sea followed necessarily (7: 443). Galileo presented the combined motions of the Earth as a simple mechanism which (if no other considerations were relevant) would produce an ebb and flow every twenty-four hours, or at least an oscillation of the seas. He found a suitable model for his theory in barges carrying fresh water to Venice: when a barge slows down, the water still has its own motion so it will pile up at the front of its container; conversely, when the barge accelerates, the water will go to the back of the container (7: 451). The seabeds are basins of water. Acceleration and deceleration are provided by the

double motion of the Earth, where the rotatory movement cycles between being added to the revolutionary movement and substracted
from it. This simple mechanism would then set the seas sloshing backwards and forwards but it would need refinement: the orientation and
configuration of seabed and shore were among the chief factors to be
taken into account in adapting the underlying mechanism to fit the
great variety of tidal phenomena observed from place to place. Then
the monthly and seasonal variations required some role for the Moon
and Sun, which Galileo supplied with his habitual ingenuity. His explanation, Galileo admitted, needed further work, yet he had no doubt
that he had provided more than the outline of a definitive theory.
Hindsight tempts us to dismiss the whole theory as an aberration
prompted by uncritical partisanship for Copernicanism, but it was a
worthwhile attempt to tackle an unsolved problem. Mistaken though it
was, few people were capable of making such an intelligent mistake;
nor would anyone before Newton offer a much better explanation. If
Newton is seen as having given the right answer, Galileo was hardly
further away from that answer than was Kepler.

Now Galileo had contracted to round off the *Dialogue* with an
argument that would ostensibly nullify this proof of the Earth's
motions, so Simplicio duly produces the argument of Urban, here
referred to as 'a most learned and eminent person'. No matter how
ingenious Galileo's explanation of the tides was, God could have
caused them in lots of other ways, perhaps ways we could never think
of (7: 488–9). Salviati, on behalf of Galileo, naturally concurs with
this enthusiastically. In a well-placed advertisement for the *Discourses*
Sagredo closes the *Dialogue* with a hope that they will meet again to
discuss motion.

Urban's argument, which was in fact a commonplace, has its attractions: no one would now claim that God uses Galileo's mechanism to
cause the tides. The argument is also a challenge to anyone who thinks
that at least some scientific theories can be proved conclusively. Its
weakness lies, nevertheless, in its generality. It does not help in the
pressing task of choosing rationally between rival scientific explanations. It is a form of instrumentalism (extended to physics as well as
astronomy) which at best allows us to prefer one set of calculating
devices to another, with the rider that none brings us towards the truth.
That is why Clavius rejected it, as we saw in chapter 2: it cuts us off
from any hope of ever discovering the true causes of anything. In this
Galileo took the same position as Clavius: in this he was close to
Aristotle's aims and methods. Whatever the shortcomings of their philosophy of science, Clavius and Galileo were right in thinking that

contemporary instrumentalism simply evaded problems that had to be faced.

Trial

The *Dialogue* was published in February 1632; by the middle of August it was suspended, though it was not referred to the Inquisition immediately. Instead the Pope appointed a special commission to examine it and the events preceding publication. Galileo was in serious trouble, but even at this stage no one could have predicted the outcome. Things could have turned out worse than they did, though one hardly likes to think about it; they could certainly have turned out better, perhaps with the book being withdrawn till corrected. But on 4 September Francesco Niccolini, the Tuscan ambassador, had an audience with the Pope and came away with the impression that the world was about to collapse. Niccolini defended Galileo bravely, but the Pope erupted. He felt he had been deceived by Galileo and his dupes, Riccardi and Ciampoli. Niccolini politely enquired whether His Holiness did not think it a good idea to let Galileo know in advance what was unsatisfactory about the book so that he could defend himself. Urban replied angrily that that was never the way of it: in these matters the Inquisition simply made its judgement and then called people in to recant; anyway, Galileo knew very well what was wrong with the book (a reference to their private conversations). Urban did point out that the appointment of a special commission was something of a favour; it would examine the book word for word, since this was the most perverse subject one could ever handle. Niccolini concluded that there was not much point in dealing directly with Urban. The only thing to do was to use intermediaries and play for time (14: 383–5). This he did with great devotion: he even braved another audience.

The 1616 decree was the touchstone for the commission, which reported early in September. But they also discovered the scruffily documented but very strong injunction described in chapter 6; naturally they suspected that Galileo had concealed it. This was something Galileo could not have foreseen. He had been convinced that, so long as he stayed within the limits laid down by the Congregation of the Index, as interpreted by Riccardi, his book would be safe. So it is worth asking: if the matter of the personal injunction had not come up, could he have got away with what he published? He would have had to answer for using the Roman *imprimatur* without permission; on that point he could probably have pleaded misunderstanding, since Riccardi was hardly blameless. He could surely have explained away the commission's complaint that the introduction he had agreed with Riccardi was in a different type and that the 'medicine of the end' (Urban's argument) was put in the mouth of a fool, Simplicio. He could, however, hardly have refuted the commission's accusation that sometimes he asserted Copernicanism absolutely, called arguments in its favour 'demonstrative' and 'necessary' and treated the opposite view as impossible. Even treating the issue as undecided, the commission noted, was not compatible with the decree of 1616 (19: 324–7).

On 15 September Niccolini learnt that, despite all his efforts, the book was to be referred to the Inquisition. Thereafter no appeal to Galileo's age and illness or to the dangers of winter travel in a time of plague were of any use. He would have to come to Rome, in chains if necessary, to stand trial. The brusqueness of the final summons was tempered by the consideration shown to Galileo when he at last arrived in February 1633: though on remand, as it were, he was allowed to stay with the Tuscan ambassador, who was solicitous towards him in every way. Even when he had to move to the Inquisition's premises he was given comfortable quarters and allowed some freedom. None of this altered the seriousness of the proceedings.

In his first interrogation and deposition on 12 April he summarized how he understood what Bellarmine had told him in 1616: Copernicus's opinion could be held suppositionally (hypothetically), but taken absolutely it could be neither defended nor held. To support this recollection he produced a copy of Bellarmine's certificate. Next came the question of whether anything else happened at the interview with Bellarmine. Galileo could remember that some Dominicans were present, though he was not sure at what stage they came, nor could he recall being given any injunction. Anyway, he had neither held nor defended Copernicanism, so he had not violated the decree. He was then informed of the stronger injunction not to hold, defend or teach the said opinion in any way whatever. All he could recall was 'hold', 'defend' and perhaps 'teach', but not 'in any way whatever'. Perhaps it had been used, but if so Bellarmine's certificate, which mentioned only 'hold' and 'defend', had made him forget it. Consistently with this perfectly credible account, Galileo said he had not asked permission to write the book since in it he was not holding, defending or teaching the forbidden opinion; presumably he was relying on the frequent disclaimers he had sprinkled throughout the text, in concert with Riccardi. But he went too far when he claimed that the book refuted Copernicus and showed that his reasons were invalid and inconclusive. He repeated this as his reason for not telling Riccardi about the injunction of

25   The Grand Duke's villa on the Pincio, from *Descrizione di Roma moderna*
(Rome, 1697).

1616 (19: 336–42). One cannot cavil at Cardinal Oreggi's report on 17
April that from the whole context of the *Dialogue* it is clear that in it
Copernicanism is defended and held. On the same day his fellow-
commissioner, the Jesuit Melchior Inchofer, gave many references to
pages where the *Dialogue* teaches Copernicanism absolutely. Inchofer

also brought in the *Letter to the Grand Duchess* as further proof that Galileo was a Copernican. His lengthy report even claimed that Galileo's principal aim was to attack Scheiner, the most recent anti-Copernican writer. This hostile reviewer had no difficulty in showing that, whatever its many disclaimers, the *Dialogue* really was a Copernican book: he saw clearly enough that Galileo had only pretended to conform to the decree of 1616. The same conclusion was reached in the much more able report of Zaccaria Pasqualigo (19: 348–60).

If Galileo persisted in claiming that his book refuted Copernicanism, things were going to be very awkward, so the Commissary General obtained permission to talk him round informally. The result was that in his second deposition on 30 April Galileo had to admit that a reader, unaware of his good intentions, would think the arguments from sunspots and the tides stronger than was intended by Galileo, who really thought them inconclusive. His excuse was the natural pleasure everyone takes in showing himself cleverer than others. He underlined this unintentionally by popping back in and offering to write additional *Dialogues* which would prove he did not hold Copernicanism. On 10 May he submitted his one real defence, Bellarmine's certificate. In a very dignified deposition he pointed out that it showed no trace of any special injunction given to him beyond what was required of all, namely not to hold or defend Copernicanism. So there had not been any reason to mention this matter to Riccardi. He had, moreover, submitted his manuscript for censorship. The flaws in his book were not the product of evil intention but of vainglory, as he had already admitted: he would do all he could to compensate for this if allowed to. Finally he asked for consideration for his poor physical health, mental stress, age and good name (19: 342–7).

By the middle of June the Pope had received the final report on the case. It first rehearsed the events of 1615 and 1616, including Lorini's complaints against the *Letter to Castelli*. Through a misreading of its own documents the propositions censured by its consultors in 1616 were now treated by the Inquisition as quotations from the *Letters on Sunspots*. Even Galileo's correspondence with German mathematicians was mentioned. Much more damningly the report took it as fact that Galileo had promised to obey the stronger injunction by the Commissary of the Inquisition under threat that otherwise proceedings would be started against him. Moving on to the licensing of the *Dialogue* the report was noncommittal, but left the impression that Galileo had no right to claim Riccardi's *imprimatur*. The rest of the report was simply a full and fair account of Galileo's own recent depositions (19: 293–7).

Galileo had to be called in again on 21 June to see whether he would persist in denying that he had held Copernicanism after the ruling of 1616. He replied stoutly that since that time he had held, and still held, the stability of the Earth and the motion of the Sun as very true and undoubted. Pressed on the point that the *Dialogue* gave rise to the contrary presumption, Galileo relied on his only possible defence: his book explained what could be advanced for one side or the other and showed that no proof was conclusive either way, so he did not hold the condemned opinion. This defence, though it could easily be holed, might have worked in a more sympathetic climate: it boiled down to claiming that he had not publicly contravened the Church's ruling. To the Commissary this must have seemed a flimsy defence against the charge of perjury, so Galileo was given two more chances to tell the truth, the second under threat of torture. Galileo stolidly repeated that he was there to obey but that he had not held Copernicanism after February 1616. One would think that this was an impasse. Yet when the Cardinals of the Inquisition passed judgement the next day this was considered a Catholic answer, presumably because of his willingness to submit (19: 361–2).

Condemnation

The actual sentence was signed by only seven of the ten cardinals present. It included the serious charge that Galileo had concealed and disobeyed the strong injunction given to him in 1616, but it also took care to show that, even on the strength of Bellarmine's certificate alone, he was guilty of defending Copernicanism and showing it as probable, a clear contravention of the decree of 1616; there was no way in which an opinion declared to be incompatible with scripture could be probable. (This is proof, if any were needed, that permission to hold Copernicanism 'hypothetically' did not mean that it could be treated as a likely conjecture.) Hence all Galileo's disclaimers about impartiality or 'chimeras' were dismissed by the judges as subterfuges.

In the sentence the judges declared him 'vehemently suspect of heresy', that is, of holding and believing a doctrine which is false and contrary to scripture and of maintaining that one may hold and defend an opinion as probable after it has been declared and defined as contrary to scripture. Abjuration would absolve him from the attached penalties, but he would be subject to formal imprisonment and obliged to say the penitential psalms weekly for three years. His book would be publicly condemned. Galileo was there to obey: he knelt and read out the abjuration under oath (19: 402–7). If he had been taxed with perjury, he could have pointed out that in free discussion he had always been willing to let his yea be yea and his nay be nay. He had done what he could to prevent the ruling of 1616; in the *Dialogue* he had tried to undo its harm while treating it with ceremonious respect. He had failed. The sentence was designed to secure what Bellarmine and the decree of 1616 had not managed: Galileo's silence and the compliance of Catholics.

To silence him real imprisonment was not necessary; control and isolation would do. He left Rome on 6 July for Siena, where he was an honoured guest of the Archbishop, Ascanio Piccolomini, who so effectively saved his sanity at this traumatic period that before long Galileo was again applying himself to mechanics. In December, perhaps because of complaints that the Archbishop's palace did not keep him sufficiently isolated, he was permitted to return to Arcetri, close to his daughters' convent, but effectively under house arrest, since he was not even allowed to go to Florence without permission, which was sometimes refused, sometimes given grudgingly. Urban's smouldering resentment died down somewhat over the years, but it never disappeared. He allowed himself to be persuaded that Galileo had not made fun of his argument, but he never saw that the argument itself could not bear the weight he was putting on it. Perhaps Urban and the Inquisition also reckoned that a man who had disregarded one injunction might offend again. Perhaps they were frightened of his powers of persuasion: they had, after all, read the *Dialogue*. Still, the awkward case was concluded without a breach with the Grand Duke. Galileo was silenced. Urban could distract himself on 29 June by the unveiling of Bernini's stupendous baldachin over the high altar in St Peter's.

The Jesuits and the Condemnation

Galileo and his friends needed some way of accounting for the traumatic shock of what had happened. They were convinced that it was the Jesuits who had brought about his condemnation out of enmity; sometimes the charge of hypocrisy was added. In that small world of easy contact, clerical and academic gossip generated reports which can easily seem to confirm this conspiracy theory. Gabriel Naudé, writing from Rome to Pierre Gassendi in April 1633, asserted roundly that Galileo was brought before the Inquisition by the machinations of Scheiner and other Jesuits who wished to destroy him (15: 88). Yet there is no evidence to substantiate this. Perhaps no delation was needed to

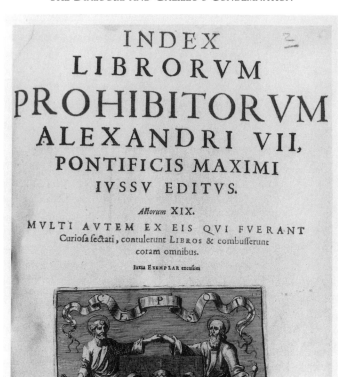

INDEX
LIBRORVM
PROHIBITORVM
ALEXANDRI VII,
PONTIFICIS MAXIMI
IVSSV EDITVS.

Actorum XIX.

MVLTI AVTEM EX EIS QVI FVERANT
Curiofa fectati, contulerunt LIBROS & combufferunt
coram omnibus.

Iuxta EXEMPLAR excufum

ROMÆ,
EX TYPOGRAPHIA REV. CAM· APOST.
Cum Priuilegio.

M DC LXVII.

26 Galileo's *Dialogue* and all books maintaining the truth of heliocentrism were listed as forbidden reading for Catholics in the *Index librorum prohibitorum.*

27   Bernini's baldachin in St Peter's, opened by Urban VIII in 1633, from
Descrizione di Roma moderna (Rome, 1697).

set in motion an inquiry, since any reader could see that the *Dialogue*
was at least suspect. Did any Jesuits work against Galileo once the case
started? Inchofer certainly made the strongest case he could against the
Dialogue, though his incompetence in astronomy at least clears him of
hypocrisy. Scheiner, the chief suspect, was very happy to write a book

to justify the condemnation after the event, a book which his superiors withheld till 1651, but his *Schadenfreude* does not make him a hypocritical crypto-Copernican. His fierce resentment may have prompted him to stir up Inquisitors against Galileo, but I do not know of evidence that he did, unless we assume that he primed Inchofer. Grassi acknowledged in September 1633 that he had been asked his opinion the year before, presumably by the commission or the Inquisition. He claimed to have done everything he could to calm those who were exasperated with Galileo and to get them to understand his arguments. Grassi's conclusion was that Galileo had brought about his own ruin: his inflated self-esteem and disregard for others made it no wonder that everyone conspired against him – a definite confirmation of the general conspiracy theory, though not specifically of Jesuit participation in it (15: 273). Grassi was no Copernican, despite Galileo's and Guiducci's wishful thinking. We have already seen that he was prepared to make trouble for Galileo over Eucharistic theology. Still, it is hard to believe that he contributed actively to the case against Galileo. Grienberger greatly enjoyed the *Dialogue*, as Evangelista Torricelli had reported in October 1632, but he did not think Copernicanism was true (14: 387). But in July 1634 Galileo had a report that Grienberger thought that if Galileo had known how to keep the goodwill of the fathers of the Roman College he would still be living gloriously, free to write whatever he wished on any topic, including the motion of the earth (16: 117). I do not know what to make of this, if we accept the report as accurate. Grienberger cannot have thought that if Galileo had never said a word against a Jesuit then he could have ignored the decree of 1616. Possibly all he meant was that Galileo should have followed the Jesuits in treating Copernicanism as a calculating device. Still, the report cannot be dismissed. Perhaps it does, after all, point a finger at Scheiner, who may have advised or egged on Inchofer: both can fairly be called partisan. Yet Galileo *had* breached the decree of 1616. An enraged Pope did not need any stirring by resentful Jesuits; it must be doubtful whether even a united defence of Galileo by the Jesuit mathematicians could have saved him.

Public Enforcement

To secure the compliance of Catholics called for an organized campaign to make clear that this had not just been a disciplinary proceeding against Galileo. Nuncios and Catholic universities were instructed to proclaim the sentence. Typical was the reply of Matthew Kellison,

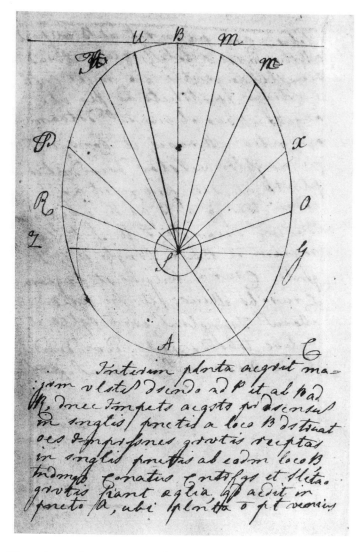

28   Notes from a lecture course of 1757 given by William Wilkinson at the English College, Douai: the College first accepted heliocentrism (and Newtonianism) only two years earlier. It was still slightly in advance of official Roman policy.

President of the English College at Douai. He has informed the chancellor and other professors of the university: so far from assenting to the fanatical opinion just condemned, they always think it should be exploded and hissed out of class. In his own seminary the paradox of a moving Earth has never been approved and never will be (19: 392–3). This was all perfectly true (apart from the prediction), heightened only by the rhetoric of loyalty. In fact, (again typically) heliocentrism was not taught as true in his college until 1755, two years before Pope Benedict XIV decided to omit from the *Index* all books advocating a moving Earth.[5] The 1616 compromise of allowing Copernicanism hypothetically did leave room for Catholic scientists to participate in scientific developments, though it led to curiosities such as an excellent annotated edition of Newton's *Principia* being prefaced by a reminder that the commentators treated Newtonianism only hypothetically.[6] In fact the ambiguities of 'hypothesis' were useful to Catholics who happened to live in countries where the Inquisition's rulings were acknowledged. The effect of Galileo's condemnation on science in Catholic countries has often been discussed: a fair generalization is that it was not as harmful as one might have expected.[7] Research is still needed into its effects on Catholic theologians and on courses of theology and philosophy in Catholic universities and seminaries. One astonishing thing is that the condemnation did not prevent Galileo himself from completing and publishing his greatest work, the *Discourses*.

9

Two New Sciences

Galileo's permission to return to Arcetri was hedged round with restrictions which could not be relaxed without the Pope's permission: he was not to invite people there nor admit callers to discussions (19: 286). That did not stop the Grand Duke paying Galileo a courtesy call of almost two hours. Reporting this on 7 March 1634 to a faithful friend in Paris, Elia Diodati, Galileo expressed pleasure that his *Dialogue* was being translated into Latin; his health was better than it had been for years, so he would be able to prepare his other researches for publication (16: 59–60). But though the Grand Duke could visit Galileo, he could not help him. On 23 March the Inquisition brusquely refused a petition to let him return to Florence itself for medical attention, instructing the Florentine Inquisitor to warn him to desist from such requests, otherwise the Inquisition would be obliged to recall him to its prison (19: 286). Archbishop Piccolomini was astonished at this: all one could do was shrug one's shoulders and be silent (16: 80). In the same letter he did his best to console Galileo for the loss of what meant most to him in the world: Maria Celeste had died at the age of thirty-three on 2 April. At the end of the month Galileo described his health as seriously threatened: the hernia (which had prompted the petition to be allowed to go to Florence) had come back, worse than ever, he had palpitations of the heart, he was immensely sad and melancholy, he was hateful to himself and continually heard his beloved daughter calling him. This was not the time for his son Vincenzio to be thinking of taking a journey, when Galileo was seriously worried by perpetual insomnia (16: 84–5). This seems to have been the lowest point of his bereavement, but he had more to do before he could follow Maria Celeste's call; he would make good his promise to Diodati.

In July he wrote encouragingly to Matthias Bernegger of Strasbourg, who had translated Galileo's little book on the geometrical compass and corresponded with him about the Moon more than twenty years earlier. Galileo carefully noted that Bernegger had undertaken the translation of the *Dialogue* into Latin without his knowledge, but when in August the Congregation of the Index got round to including the *Dialogue* in its list of forbidden books, the expected prohibition made no difference to his willingness to see it have a wider circulation (16: 111–12; 19: 416). (A couple of years later he was very happy to learn that his *Letter to the Grand Duchess* was at last in print, both in Italian and in Diodati's Latin translation (16: 445).) His failing eyesight and chronic ill health made it imperative not merely to disseminate more widely what he had already published in Italian but also to prepare for publication what he knew would be his master-work, his two new sciences, especially the science of motion. His correspondence frequently shows him in reflective mood. He often rehearsed the story of his sufferings; his conscience was clear; if friends like Peiresc could not obtain better treatment for him that was because authority can only pardon where there has been a fault (16: 215). But his melancholy did not keep him from work. Indeed, as he reported in November 1634, new ideas kept his mind churning, so that there was more to completing his final book than mere editing of work already accomplished. In the same letter he responded to a request to compare himself with Kepler, who had died in 1630:

> I have always esteemed Kepler for his free (perhaps too free) and subtle mind, but my way of philosophizing is very different from his; it may be that writing on the same topics (celestial motions, nothing else) we have occasionally hit on similar ideas, though very few, so that we have assigned the same true reason for the same effect; but that would not be true of one percent of my thoughts. (16: 163)

In suggesting that Kepler's mind was 'too free' Galileo was not hinting at any differences in attitudes to theology or ecclesiastical authority. What put Galileo off, as we have already noticed more than once, was Kepler's fascination with Pythagorean harmonies. In failing to appreciate the tremendous theoretical advance which Kepler's strange preoccupations and indefatigable persistence had made possible Galileo had the company of most astronomers at that time. If he had managed to

29   Nicolas Fabri de Peiresc, pupil and supporter of Galileo, from Isaac
Bullart, *Académie des Sciences et des Arts*, tome 2, livre 2 (Amsterdam, 1682).

seize on the lasting achievements of Kepler and incorporate them in his
own programme our admiration for his scientific judgement would
undoubtedly be greater. But it is no great blemish on his scientific
reputation that he was fallible.

The ban on entertaining visitors was not interpreted too strictly. In December 1635 Thomas Hobbes dropped in with the information that the *Dialogue* had been translated into English (though it was a different translation, Thomas Salusbury's, that was first published and that was not till 1661 (16: 355)).

It was relatively easy for Galileo to disown responsibility for translations of his published works, since everyone knew that such copyright as authors had was not effective where the jurisdiction of the licenser was not acknowledged. But he had to give serious thought about where he was to publish his final book. Venice, still a great centre of printing, was the obvious choice, but it had to be ruled out when the local Inquisitor showed Micanzio that Galileo was not allowed to publish anything, not even the Lord's Prayer or the Creed, as Micanzio put it (16: 209, 230). It was elementary prudence for Galileo to prepare some sort of story which would allow the Inquisition to sit quietly without contesting his pretence that a manuscript of his had been published without his knowledge. Given the very free circulation of manuscript material at the time such a story was easy to set up. It was a neat touch to put it under the protection of the French ambassador to Rome, François de Noailles, a former pupil of Galileo's who had done his very best to end, or at least ameliorate, Galileo's confinement at Arcetri. (He was the one who managed to persuade the Pope that Galileo had never had the slightest thought of mocking him in the conclusion of the *Dialogue* (16: 450).) So when the *Discourses* were published in 1638 a dedicatory letter by Galileo explained that he had given a manuscript copy to Noailles when they met so that his researches, preserved in manuscript, would not perish without trace. He was thinking of sending copies to other countries when he heard unexpectedly that the Elzevirs had the work in the press. This startling news he could only take as a compliment, a sign of friendship from Noailles.

The story was plausible enough since it was laced with a dash of truth. (Galileo was allowed to travel to Poggibonsi to meet Noailles in October 1636 (16: 507).) Its chief purpose, however, was to signal that Galileo was not engaged in subversive activity which the Inquisition would have to notice. In this it was successful. It would have taken a very prescient Inquisitor to divine that these mathematical discourses were at least as subversive of traditional understanding of nature as anything Galileo had already published. Naturally Galileo did not present his new science of motion as essential to his Copernicanism. It is, in any case, very debatable how much a firm belief in Copernicanism motivated his Paduan studies of falling bodies and projectiles, so there was no glaring oddity in presenting his discoveries in their own right, though had the *Dialogue* proved acceptable then Galileo would undoubtedly have brought home to his public that the *Discourses* provided an understanding of how things move on an Earth that is itself moving. The whole second day of the *Dialogue* looked forward to the *Discourses*, but in the *Discourses* the one thing that Galileo could not do was look back.

The true story of how Galileo set about finding a publisher is long and complicated. It was not sufficient to publish abroad. A publisher had to be found who was not only competent but also invulnerable to hostile machinations. At times during the protracted correspondence with friends scattered in various parts of Europe whole countries were deemed unsafe from Jesuit influence. The presence of Scheiner alone could rule out a city like Vienna (17: 130). Paranoiac though this may sound, it rested on the quite sensible fear that he might be given the job of examining the manuscript or at least that he might prevent its publication there or elsewhere (presumably by putting it about that Galileo was forbidden to publish anything). At one time when it looked as though Cardinal Dietrichstein, who had his own press, would provide a safe method of publication, Galileo sent what he had already written. The death of the Cardinal ended that hope. It was sometimes hard for Galileo to keep hope alive. At the end of 1636 months of silence from Germany made him think that half of his book had been irretrievably lost:

> Unhappy land of ours, ruled by a fixed resolve to want to exterminate all novelties, in particular in the sciences, as though everything knowable is already known. (16: 361)

Eventually Galileo took up the suggestion of Diodati, who arranged for Lodewijk Elzevir of the famous Leyden firm to visit him at Arcetri in May 1636. Before very long Elzevir had the copy for the first three days of the work (16: 436, 510). From then on preparations for publication went as smoothly as could be expected, given the difficulties of communication and Elzevir's extensive travelling before he returned home with most of the manuscript at the end of the year. The completion of the fourth day on projectiles was one of Galileo's greatest triumphs. At the end of 1636 he was still making new discoveries:

> And although in this part too I open the entry for speculative minds to spread themselves over an immense field, I should still like to express myself a bit more fully. But I find how much old age lessens the liveliness and speed of thinking, as I struggle to understand quite a lot of things I discovered and proved when I was younger. (16: 524)

He had finished the work by February 1637. In July he lost the sight of his right eye (17: 126). Nevertheless, in November he reported that he had discovered a libration of the Moon (17: 214–15). By the end of the year he was completely blind. He dictated a letter to give Diodati the sad news: the universe, which by his marvellous observations and clear demonstrations had been expanded a thousandfold from what had been visible to the wise of all ages, was now for him confined to his own person (17: 247). At the beginning of 1638 it seemed to Galileo that at last he might have his freedom restored: Castelli sent him a suitably phrased petition to use (17: 255). Permission to go to Florence for medical treatment was given, but only grudgingly after an unannounced visit in February 1638 by a local Inquisitor, accompanied by a doctor. Galileo's blindness and pitiable state of health were all too evident. The Inquisitor satisfied himself that there was no danger that Galileo would disseminate the condemned opinion of the Earth's motion. Blindness had ended his studies, apart from occasional listening to passages from books. Cardinal Francesco Barberini could be sure the Pope could safely let Galileo go to Florence. If Galileo did try anything, a sharp warning would be enough to bring him into line (17: 290). Galileo, then, was no longer a threat. Nevertheless, when the permission was granted it was conditional on Galileo's not discussing the condemned opinion: it was the Pope himself who insisted on this condition (17: 310–11).

The same preoccupation was evident later in the year when the question was whether Galileo could receive an envoy to discuss the problem of longitude at sea. Galileo had never given up hope of seeing his own method implemented by a seafaring nation. For years he had engaged in extensive correspondence with representatives of the States-General. As a sign of their gratitude the Dutch sent Galileo a valuable gold chain. In July 1638 Cardinal Francesco Barberini advised the Florentine Inquisitor that, if the envoy was heretical or from a heretical city, then Galileo should be told not to admit him to discussion. A Catholic from a Catholic city could be admitted, provided the Earth's motion was not discussed. Barberini was sceptical that Galileo had made an instrument which showed how to navigate by the longitude of the pole, but his loose grasp of the problem did not lessen his patriotic sense that the Grand Duke should not allow any such invention to fall into foreign hands (17: 356). When Galileo prudently refused the handsome gift Barberini promptly instructed the Florentine Inquisitor to let him know that this sign of his great piety had been very well received by the Cardinal Inquisitors (19: 366).

The Florentine Inquisitor had not been far wrong when he had reported that Galileo was failing badly. He still hoped to complete supplementary days to follow those in the *Discourses*. Almost to the very end he carried on important correspondence by dictation. But his chief work had been accomplished. The *Discourses* were published in the spring or early summer of 1638. Galileo himself received no advance copies and was kept waiting before he received a copy of his own. It arrived when he could only handle it and have it read to him. But he knew that once it reached other speculative minds it would open the way to a succession of more marvellous discoveries and proofs (8: 267). That was the most sober of the claims he made to intellectual glory: in the many controversies surrounding this fascinating man it is a claim that has withstood all attempts to erode it – glory enough, even without his other great achievements.

The Discourses

Although Galileo was not happy with the title which the publisher gave to his book, *Discourses and Mathematical Demonstrations concerning Two New Sciences* (*Discorsi e dimostrazioni matenatiche intorno a due nuove scienze*), it seems apt enough: the two new sciences deal with the strength and resistance of materials and with motion. Galileo brings together the same three participants to discuss topics at their leisure for four days, but it would be misleading to call all the books (or days) dialogues. The third and fourth days are dialogues only intermittently. They are seminars conducted by Salviati to present Galileo's most important discoveries: the Italian dialogue serves to introduce or explain readings from a Latin treatise on how bodies move. The convention that Salviati reads these expositions from manuscripts written by Galileo serves its purpose, but the juxtaposition of Latin lecture and Italian dialogue is often awkward. Still Galileo was clearly right to present his teaching on motion as rigorously as possible in the last two days, so written texts had to be fitted into the dialogue somehow; the fact that they were in Latin made them available to foreign readers. At the same time he did not want to forgo the lively conversational pieces which often contribute significantly to the explanation of his theories as well as providing some relief from the rigours of geometry. Carugo makes the attractive suggestion that Galileo had not forgotten the needs of engineers, so that the *Discourses* can also be seen as a manual for them, catering for building engineering on the second day, and hydraulic and military engineering on the third and fourth days respectively.[1] One can also detect Galileo's concern to include as much as

DISCOURSES

CONCERNING

Hijilion

Two New Sciences

RELATING TO

Mechanicks and Local Motion,

IN

FOUR DIALOGUES.

I. Of the Refiftance of Solids againft Fraction.	III. Of Local Motion, *viz.* Equable, and naturally Accelerate.
II. Of the Caufe of their Coherence.	IV. Of Violent Motion, or of PROJECTS.

By GALILEO GALILEI,

Chief Philofopher *and* Mathematician *to the Grand Duke of* TUSCANY.

With an APPENDIX concerning the Center of Gravity of
SOLID BODIES.

Done into *Englifh* from the *Italian*,
By THO. WESTON, *late Mafter, and now publifh'd by* JOHN WESTON,
prefent Mafter, of the Academy at Greenwich.

LONDON:

Printed for J. HOOKE, at the *Flower-de-Luce*, over-againft St. *Dunftan's*
Church in *Fleet-ftreet.* M.DCC.XXX.

30   The title-page of Thomas Weston's translation of the *Discourses*
(London, 1730).

possible of his unpublished work of five decades: that partly explains
why the first day, which starts off by discussing the strength of materials, meanders off into digressions to cover such topics as the musical
studies carried out by Galileo and his father more than fifty years

earlier. Even Galileo could not pretend that his theorems on the centre of gravity of solids would slot into anything resembling a dialogue, but he naturally wished to preserve the youthful work which had laid the foundations of his career, so he included it as an appendix. The topics of the discourses did not all lend themselves easily to the conversational approach of the *Dialogue* and in any case there was no point in Galileo's taking unnecessary risks by making fun of Aristotelians, or anyone else. He did allow himself two or three complaints, including one about Grassi, who had insinuated that Galileo denied divine providence, but in general he stuck to his appointed task (8: 72).

In this more sober book Simplicio is a serious participant, something which surprised Castelli, who found the Simplicio of the *Dialogue* truer to life (18: 26). It is, however, noticeable that the more mathematical the treatment becomes the less there is for Simplicio to contribute, despite Salviati's endeavours to explain everything as he goes along. In other words, Galileo recognized that his most important work had to be addressed to those who were already equipped to follow it. If one had to pick a moment when basic science began to elude the grasp of the educated public of Europe, one could do worse than fix it at the beginning of the third day of these discourses, when Galileo begins his mathematical treatment of uniform motion. Even a Sagredo, whom Galileo called his 'idol' in a letter of April 1636 (16: 414), would have been stretched. The first two days could be followed by Simplicio to some extent, partly because the first of Galileo's two new sciences seemed fairly easy to grasp, at least in outline, and partly because the first day contained a good deal of introductory material on motion. Strictly speaking this treatment of motion could just as well have gone into the third day, except that it was not cast in the form of theorems and propositions: it was really a revision of Galileo's earliest writings on motion which we saw in chapter 3. For that reason it was nearer to what an Aristotelian like Simplicio could follow.

The First Day

We saw in chapter 4 how Galileo began the *Discourses* by saying how much could be learned from the skilled foremen of the Venetian Arsenal. That passage not only set the scene in the same city as his condemned *Dialogue*, a clear signal that he was continuing his earlier contributions to science, but it also led naturally into questions of how scale affects the constructibility of structures. Could there be a ship or a building or an animal of any size we please to specify or do the natural strength and resistance of materials set limits to what is possible? Galileo undertook to settle these and related questions by geometrical demonstrations (8: 54): that is what makes his treatment into a new science. Most of these questions had to wait until the second day, since the conversation soon took Salviati into discussion of the vacuum; the continuum; the divisibility of matter and attendant puzzles about whether things are made of an infinite number of particles and vacua; and what 'infinite' means. Descartes was again exasperated by Galileo's wayward readiness to dart down every promising sidetrack, but this first day is a good deal nearer to Galileo's (and other scientists') working methods than the discoveries tidied up for public display in the third and fourth days. Descartes also thought it a weakness that Galileo offered only reasons for some particular effects without reference to first causes in nature (17: 387). This supposed weakness has generally been seen as the most judicious part of all Galileo's scientific strategy: the search for causes would come only after a proper grasp of the effects had been reached. Galileo, unlike Descartes, was not trying to put the whole of knowledge on a firm foundation: he was just opening up a couple of new sciences of great importance.

Discussion about whether there could be motion in a vacuum introduces the revised version of Galileo's early 'Archimedean' account of motion. It is remarkable how much Galileo is able to salvage from his first attempts. In repeating the arguments which show that Aristotle was mistaken in thinking that the speed of falling bodies is proportional to their weights and inversely proportional to the resistance of the medium Galileo adds only a few refinements which he had known for decades. He allows, for instance, that in a fall of 200 feet a ball of 100 pounds will reach the ground two finger-breadths ahead of a one-pound ball. But the point still stands that, according to Aristotle, the small ball should only have fallen a couple of feet by that stage. After further discussion, reminiscent of his early work on the resistance of the medium and his book on floating bodies, Galileo concludes baldly that in a medium where there was no resistance all bodies would fall with the same speed, a marked contrast to his early view that they would fall with speeds proper to their specific gravities (8: 116). He had long since abandoned the opinion that acceleration was only a transient stage at the beginning of motion until the body reached its proper natural uniform speed. It is on the third day that he sets out his mathematical account of accelerated motion. But in this introductory account he does recognize that the resistance of the medium will eventually impose a terminal speed on a body which falls far enough (8: 119). His concern here is to remove the confusions which had misled him and still misled Simplicio, confusions caused by the effects of different media on falling bodies. So it is natural that Salviati should tease out the implications of all this for Simplicio. He grants that we cannot test falling bodies in a vacuum, so we have to extrapolate from what happens in a subtle medium like air. But this extrapolation or simplification makes good sense of what happens because we can estimate what difference the resistance of the air will make to falling bodies of very different specific gravities. It need no longer puzzle us that an ebony ball falling through air from a considerable height loses only one-thousandth of its natural speed to air resistance since it is a thousand times heavier than air, while a bladder loses a quarter and so reaches the ground noticeably later. Once this is grasped, Simplicio should be able to see that in a vacuum both would fall together (8: 120).

Salviati emphasizes how new and, at first sight, improbable it seems that weight should have nothing to do with speed of fall, so no corroborating experience or reason should be left out. Sagredo loyally adds that many other propositions of Salviati are so remote from commonly received opinions and teachings that their publication would stir up many adversaries, given that it is human nature for men to look sourly on others who discover truths or expose falsehoods they have missed. Such people employ the unpopular label 'innovators of teaching' to cut knots they cannot loosen and use subterranean mines to demolish buildings which have been constructed normally by patient workmen. He and Simplicio are not like that: they are satisfied with the reasons and proofs given already, but if Salviati has more impressive ones they will be very glad to hear them.

So Salviati explains the difficulty of measuring fall accurately: in a long fall the air interferes too much with light bodies; in a short one it is difficult to decide whether heavy and light bodies hit the ground simultaneously or only very nearly so. That is why he used inclined planes to slow down their motion and make repeated observations convenient. Better still for this purpose are pendulums. Take two such pendulums with strings of equal length, one with a cork bob, the other with a lead one; pull them back an equal distance from the perpendicular and release them at the same instant. They will swing back and forth in perfect unison: even in a hundred or a thousand swings the heavier bob will not get ahead by a single moment. The air will lessen the amplitude of the cork's vibrations, but it does not affect their frequency. This confuses Simplicio at first, but Salviati eventually convinces him that when the lead and the cork fall through the same arcs they must be travelling at the same speed, which is the point of the experiment. Simplicio declares that if he had his time over again he would start by studying mathematics (8: 128–34).

Those less docile and more competent than Simplicio were not so easily satisfied, though it was not Aristotelian ideas of heavy bodies falling faster than worried them. MacLachlan, in a careful and sympathetic reconstruction of Galileo's published work on pendulums, instances Marin Mersenne, a French Minim friar, who was Europe's unofficial secretary for the sharing of scientific information. In 1639 Mersenne found that the size of the arc matters: a bob takes longer to swing to the vertical through 90 degrees than it does through ten, so amplitude does affect frequency. More generally, MacLachan concludes that this particular experiment is imaginary: if both pendulums are started from 50 or 60 degrees, they would get out of step by several counts in a hundred.

MacLachlan is by no means suggesting that all Galileo's reported observations of pendulums (or other things) are imaginary. He points out that Galileo frequently used words like 'almost' to cover the gap between the observed and the ideal case. Moreover, however imaginary or idealized the presentation, Galileo did, as MacLachlan admiringly points out, correctly conclude that the speed of the lead and the cork bobs are the same through the same arcs in the first quarter-oscillation to the vertical.[2]

This airing of a topic which really belonged to the last two days fitted in with Galileo's relaxed attitude to his agenda. Before long he lets pendulums suggest problems in musical theory, a natural enough digression to topics he had discussed at home with his father fifty years earlier, while hanging weights from strings to put the strings under tension. What Vincenzio would have found novel in Galileo's approach is not anything in musical theory, but the way a basic physics of sound is outlined almost casually. Sagredo and Salviati conclude the first day with the cheerful acknowledgement that they have hardly touched the topic they set out to discuss (8: 150).

The Second Day

Next morning the participants are all a bit more disciplined. Salviati romps through an exposition of the breaking strength of materials which leaves Simplicio testifying that, while logic may be useful for checking reasoning, it does not approach geometry's acuteness in preparing the mind for discoveries (8: 175). He is docile enough to follow a discussion about a semi-parabolic cut which would lighten a wooden beam by a third and leave it equally resistant to fracture throughout. This leads Salviati to describe an easy way of drawing parabolas, reminiscent of Guidobaldo's inked ball, though Salviati merely projects a

31   On the breaking strength of beams: an illustration from Thomas
Weston's translation of the *Discourses* (London, 1730).

(slightly heated) ball onto the surface of an almost vertical mirror. The
path it traces is clear sensory evidence that projectiles move in a parabola. Another simple way is to hang a light chain from two nails. In
fact, this catenary curve would be only an approximation to a parabola,
but Galileo realized that it would serve more than adequately for the
marine carpenters whom he had in mind (8: 180–6, 310).

The digressions of the first day were by no means as pointless as
might have appeared: speculations about atoms and tiny void spaces
held out hope of a new science of materials, based on a revision and
development of ancient ideas. Galileo had given plentiful hints about
how one might find new ways of handling the seemingly intractable
paradoxes of the divisibility of matter into parts that must themselves
be either finite or infinite in number and of some size or none. But it
was not such speculations that justify his claim to have founded a new
science of materials: the claim rested on the demonstrations of the
second day. It is a claim that is generally accepted. In a key discussion
leading up to his first proposition of the second day he was not content
with the common knowledge that a wooden beam which, when vertical, sustains a great load hanging from it, can, when horizontal, be
broken by a much smaller load placed at one end when the other is
fixed into a wall. His proposition related the resistance to fracture in the

two cases to the length of the beam and its cross-section. The rest of the second day developed similar propositions and corollaries. Now even at the time there were queries about whether Galileo had succeeded in proving his first proposition; before the end of the century Galileo's treatment could be seen to be over-simplified. But this still leaves intact his claim to have founded a new science. Historians can retrieve this or that instance where a Guidobaldo was clearer than Galileo. They do not expect to find that anyone before Galileo presented the science of the strength and resistance of materials with the same comprehensiveness and confidence in reading the book of nature. Even if a hitherto un-known manuscript were to come to light, it would not alter the fact that it was Galileo who founded this new science in the public domain and has since been given due credit by those who have built on and cor-rected his work.[3]

The Third Day

The third day begins abruptly with the introduction to a Latin treatise on local motion. It may be a sign of the difficult conditions under which the work was put together that Galileo did not introduce it with a few lines of patter in Italian from Salviati and Sagredo. The opening words are a fairly sober claim to have founded a very new science on a very old subject. The books on motion by philosophers are neither few nor small – perhaps Galileo was thinking back to his Pisan lectureship, when Buonamici's massive work was published – but Galileo finds they have not noticed, much less demonstrated, several important fea-tures of motion. Slighter aspects such as the continuous acceleration of falling bodies have been noted, but the proportion which rules that acceleration has not been given. No one, as far as Galileo knows, has shown that the spaces covered in equal times by a body falling from rest are as the odd numbers from unity. That projectiles trace a curved line was known, but no one has shown that it is parabolic. These and many other things worth knowing will be demonstrated and an entry will be opened to a vast and excellent science, of which his work will provide the elements and in which minds more acute than his will penetrate the more hidden recesses.

The treatise will have three parts: (1) on uniform motion; (2) on naturally accelerated motion; (3) on violent motion, or projectiles. The first two are dealt with in the course of the third day, while the fourth day is reserved for projectiles.

Uniform motion is dealt with briskly. It is defined as motion in which the distances travelled by the moving object during any equal intervals of time are equal to each other. There follow four axioms and six theorems or propositions. But it is with naturally accelerated motion that things become interesting, though again this part begins awkwardly with a Latin introduction, followed by discussion in Italian.

Galileo's treatment of naturally accelerated motion is meant to apply to the real world, to the way bodies actually fall. After protracted intellectual efforts he is confident he has found the way nature actually works, because what he has discovered fits perfectly with what our senses show us. The clue which led him was nature itself, which always uses the simplest and easiest means (though the discovery of these means had not been a simple or easy process). Once he had noticed that a stone falling from rest accelerates continuously (receives new increments of speed, as he puts it), why might he not believe that such additions are made according to the simplest possible rule? In uniform motion, as we saw, equal distances are covered in equal times. So motion will be uniformly and continuously accelerated when in all equal times it receives equal increments of speed: the amount of speed acquired in the first two time-intervals will be double that of the first alone; that acquired in the first three will be triple that of the first, and so on. So the increment of speed is proportional to the increment of time (8: 197–8).

It is at this stage that we see why Galileo retained the dialogue form: he needed opportunities to remove difficulties and to introduce his mathematical exposition. So Sagredo immediately doubts whether Galileo's definition fits the way bodies actually fall, since near the start of the fall the body's speed would be incredibly low, perhaps a mile a year, or even a mile in a thousand years. It is not incredible, if you think about it, replies Salviati, who is too polite to remind Sagredo that he explained this very point in the first day of the *Dialogue* (7: 46): if you throw a ball upwards it will slow up gradually, before beginning to drop; in fact, it must pass through every degree of slowness as it slows up. Similarly, a body falling from rest, that is, from infinite slowness, cannot just jump to a noticeable speed without first going through all degrees of slowness. Not that the body lingers, as it were, at any degree of speed: time can always be sufficiently divided to correspond to any number of degrees of speed or, as Salviati says, to the infinite degrees of diminished velocity. In his first writings on motion Galileo had pretty well grasped this point (1: 314–15); so here he can draw on (and clarify) some of his early insights to make things easier for Sagredo and Simplicio (8: 199–201). Another way of doing this is to have them repeat his own early mistakes or confusions.

One such plausible mistake was the version of the impetus theory that we saw in chapter 3, whereby when a body is thrown up vertically it has still not used up all the imparted impetus when it begins to fall back to earth. Sagredo puts it forward and strengthens it in answer to a sensible objection from Simplicio. Just when one is expecting Salviati to tidy the matter up in a sentence or two, he cuts across the discussion with a surprising intervention. It does not seem to him the opportune moment to enter into investigation of the *causes* of natural motion. At present our author is merely investigating and demonstrating some of the *properties* of accelerated motion, whatever the causes of the acceleration may be. He repeats what he means by accelerated motion: one in which there are equal additions of speeds in equal times. Now if the properties which he demonstrates are found to belong to the accelerated fall of heavy things, then we can take it that the definition does apply to their motion and that their acceleration increases as the time and duration of their motion (8: 202).

This famous declaration of policy has had many things read into it. It may at first sight seem to advocate an approach to physics very like the instrumentalist approach to astronomy. Against that interpretation is the fact that Galileo definitely wants his definition and what follows from it to apply to the way bodies actually fall and he expressly allows that the search for causes must be taken up at a suitable time. Here it will be sufficient to notice his strategy: whether or not we can arrive at the true causes of accelerated motion, no one has so far done so. The way ahead must lie through an understanding of that motion's properties (accidents, as Galileo calls them, in contrast to substance). Time enough to search for causes when the properties are understood (8: 202). This means, in practice, that Galileo is going to confine himself largely to what would later be called kinematics. This was surely a sensible strategy. Whether it implies a whole philosophy of science, or view of scientific method, entirely different from Aristotle's search for causes is a distinct question.

Neither Sagredo nor Simplicio jibs at this programme. Sagredo offers a clarification of Galileo's definition, which he thinks is equivalent in meaning: accelerated motion is that in which speed increases in proportion to the space passed through. Salviati is glad of the chance to dispel this confusion, which he and our author once shared, a clear reference to Galileo's proposal of this principle in his letter to Sarpi of 1604. Galileo's attempt to show what is wrong with this specious definition is not entirely satisfactory, but the more than satisfied Sagredo gives one of his little speeches saying that Salviati makes these things so easy that people appreciate them less. This is Salviati's cue for a sermon: to receive disdain rather than thanks when one succinctly exposes the fallacies in propositions which have been held true by all is bearable. But it is very displeasing and annoying when some people, who think they are anybody's equal in such studies, accept as true conclusions which another then easily shows to be false. This comes from a desire to maintain inveterate errors rather than accept newly discovered truths, a desire which sometimes leads them to write against truths which they know in their hearts to be true, solely in order to keep down another's reputation in popular and unthinking estimation (8: 203–4). It was hard for Galileo (and his friends) to do without the consolation of this conspiracy theory, yet bad faith, whether by itself or accompanied by malice, seems an incomplete explanation for lack of progress in science. Still, once Galileo has got that off his chest he is ready to get down to business.

With the definition accepted, Salviati notes that the author makes a single assumption: the speeds of a body down differently inclined planes are equal when the height of the planes is equal, the speed being that acquired in fall through the vertical height of the planes. This seems reasonable to Sagredo, provided the planes are perfectly smooth and the moving body is perfectly round. Salviati goes further: he can make it approach a necessary proof by appealing to experience. What he has in mind is the way pendulums swing. He shows that the swing will take the bob (almost) to the same height from which it was released and that this pattern is repeated even when the swing beyond the perpendicular is shortened by a protruding nail which catches the string: so the moments of speed in all such arcs are equal to the moments of speed acquired in the fall to the perpendicular. It is, of course, one thing to swing a pendulum bob and quite another to get a ball to run down one slope and then up another, so for the moment Salviati takes the assumption about inclined planes simply as a postulate: its truth will be established when the inferences drawn from it are found to agree with experience.

The participants are now ready to move on to the exposition of Galileo's theorems. A sample or two will give a glimpse of how Galileo laid the foundations of his new science of motion. The first theorem is a legacy of medieval discussions of latitude of forms and is, in fact, known as 'the Merton rule', since it was first discovered at Merton College, Oxford, in the fourteenth century. This 'mean speed rule', or more accurately 'mean degree rule', states that the distance covered by a uniformly accelerated motion in a given time is the same as would be covered in that time by a uniform motion at the mean speed of the accelerated motion. The diagram in Galileo's proof looks very similar to one given by Nicole Oresme in the fourteenth century. Yet, impressive though that and similar medieval work is, it did not lead anyone before Galileo to apply it seriously to actual motions. (Domingo de Soto, who did recognize that free fall was an instance of uniformly accelerated motion, showed indeed, though only incidentally, a clear grasp of the consequences of medieval analyses, but he was content to leave it at that.)[4] It is true that Galileo's published results of experiments often seem idealized versions of whatever measurements he had managed to make. But no matter how much he cleaned up his results for publication, it cannot be doubted that he had made serious attempts to make accurate measurements. That alone is sufficient to mark off his work from the qualitative speculations of those who are sometimes allotted the role of his precursors. As we shall see immediately, he did much more than develop a few brilliant medieval ideas – and it is still debated how influential those ideas were in his own actual work at Padua.

The next theorem is the times-squared law: if a body falls from rest with uniformly accelerated motion, the spaces it covers in whatever times are as the squares of the times (8: 209). Simplicio has difficulty with the author's geometrical proof but finds Sagredo's exposition more helpful, yet none of this shows that accelerated motion is really like this. Can he not be given an example of one of those experiences which were promised earlier? This earns him a compliment as a good scientist who has made a very reasonable request. All sciences such as perspective, mechanics, astronomy and music, which apply mathematical proofs to natural conclusions, confirm their principles with sensory experiences. Naturally the author has not neglected such experiences and Salviati embarks on the classic description of experimenting with inclined planes.

First the apparatus is described very carefully: one can understand why Crew and De Salvio slip into textbook language in their translation: 'A piece of wooden moulding was taken.' This seems an archetypal experiment. The piece of wood was about 24 feet long, 1 foot wide and three fingers deep. A very straight channel little wider than a finger-breadth was cut in its depth and, to make it smooth and polished, vellum was glued in it. Then a well-rounded and polished ball of hardest bronze was rolled down it. When the board was raised a full run was timed and all runs were found to be completed in the same time, give or take a tenth of a pulse-beat. Next a quarter-run was found to take precisely half the time of a full run. This satisfactory finding was repeated no matter what fractions of a run were tried and no matter what the inclination of the plane: the spaces traversed were always as the squares of the times.

258 $GALIL\cancel{A}EUS$'s Dial. III.

Momentum subduple of the greateft *Momentum* of the Velocity of the accelerated Motion. *Q. E. D.*

P R O P. II. T H E O R. II.

If a Moveable defcend from Reft, with a Motion uniformly accelerate, the Spaces paffed by it in any Times whatfoever, are to each other in a *duplicate* Proportion of the fame Times, *i. e.* as the Squares of the fame Times.

Let the Line AB reprefent a Length of Time from any firft Inftant A, wherein take any two Times AD and AE;

and let H I be a Line in which the Moveable from the Point H, as the firft Beginning of Motion defcends, accelerating

32   Falling bodies, from Thomas Weston's translation of the *Discourses* (London, 1730).

The times had to be measured by weighing water. Water flowed in a thin jet from a narrow pipe let into the bottom of a large container. The water was collected in a glass during each run and weighed on an accurate balance, the differences and ratios of the weights giving the differences and ratios of the times. There was no notable deviation in results (8: 212–13).

Simplicio spoke for many modern scholars when he said: 'It would have given me great satisfaction to find myself present at such experiences.' He, however, was prepared to accept Salviati's word for the results: it has been more problematic for historians. The difficulty is that the results seem too good to be true. The method of timing seems crude and the conditions of experiment remote from the theoretical ideal. Yet when Settle reconstructed Galileo's experiment he reported favourably.[5] Even the water-clock has been found to be nearly as accurate as Galileo claimed, though it should be remembered that he was working in proportions and not giving absolute times. Naylor, too, has no doubt that 'Galileo was a very capable experimental scientist who appreciated the usefulness of precise observation.'[6] But he thinks that Galileo can never have performed the experiment with parchment glued to the groove, since (if he was able to time runs as accurately as he claimed) he would have found that such a lining actually slows the ball down: this is because the inevitable joins every two and a half feet or so offset the advantage of extra smoothness. Naylor does not believe that Galileo could have confirmed his theory with the accuracy claimed. As described in the *Discourses* the experiment represents an ideal.[7]

Whether or not the experiments as described are idealized, Galileo was able to use his work on inclined planes to clear up all sorts of things that had puzzled him as well as others. His first attempt decades earlier to say how bodies fall along inclined planes not only did not fit the facts, it was misconceived. Now he can prove that when a body falls along an inclined plane and through a vertical of the same height the times (not the speeds) of descent will be to each other as the length of the inclined plane and the vertical (8: 215). In fact, once he had understood properly his own times-squared law, he was well on the way to the exposition of the theorems on inclined planes that make up the bulk of the third day of the *Discourses*. He was also able to incorporate findings which predated his grasp of uniformly accelerated motion, such as the theorem about descent along chords of circles, which we saw him proposing to Guidobaldo in 1602. His work on times of quickest descent also figured in the *Discourses*. Galileo correctly saw that the path of descent through the arc of a circle was quicker than that through the subtended chord, but he did not associate his preliminary study of the cycloid with the problem of quickest descent.

Even a cursory glance at the third day of the *Discourses* must also take in what Galileo said about inertia. In a Latin scholium he remarks that any velocity found in a moving body is of its very nature indelibly impressed on it, so long as external causes of acceleration and retardation are removed, a condition fulfilled only on horizontal planes. The question he first raised decades earlier, whether motion on a horizontal plane is perpetual, is now answered affirmatively (8: 243). Since for Galileo a horizontal plane is one which never takes the body further away from the centre of the Earth, one could see this as a law of circular inertia. One could also say that for practical purposes he was treating the plane as a straight line rather than the arc of a circle. But it seems unnecessary to claim that he fully anticipated Newton's first law of motion, even if Newton himself was happy to allow that he had. It seems sufficient to accept Galileo's own claim, made through Sagredo at the end of the third day, that our Academician had fulfilled his promise to provide a new science of a most ancient subject. Even more to the point is Salviati's assessment that for the first time a door has been opened to a new subject for contemplation, full of infinite and wonderful conclusions, which in future times can exercise other minds (8: 267–8). And there was still the study of projectiles to come.

The Fourth Day

After dealing with uniform motion and with uniformly accelerated motion Salviati can now turn to motion in which they combine. That there could be such composite motion, with neither motion interfering with the other, seemed impossible to Aristotelians. But it was one of Galileo's most significant ideas: he used it in propositions about the composition of motions in the third day and it gave him his explanation of the motion of projectiles, the topic of the fourth day. He had already said that uniform motion on a horizontal plane would be perpetual, if the plane were infinite. But if the plane came to an end, the moving body would drop off; to its uniform motion would be added the accelerated motion of a falling body. The body moving with this composite motion would trace a semi-parabola. This is the first and most important theorem of the fourth day.

Even Sagredo is less than confident that his understanding of conic sections will carry him through this topic, while Simplicio is resigned to accepting Salviati's assertions on faith, so Salviati gives a brisk tutorial on what Simplicio needs to understand about parabolas. Simplicio, who has been trying to catch up on Euclid overnight, finds that Salviati's explanation takes a lot for granted. Salviati politely elucidates an elementary point which is troubling Simplicio, but first says that all

DE MOTV

Proiectorum.

LIBER SECVNDVS.

ROIECTA nunc, bellorumq; minas, atque arcium tormenta dicemus: Supremus hic laborum Galilei fructus, suprema etiam gloria. Ostendit Galileus in libro de Motu Proiectorum, quod si mobile aliquod à plano horizontali a b decidat, impetu prius horizontaliter concepto, parabolam aliquam, vt b c. casu suo designabit. Verum est; dummodo linea a b quæ est directio proiectionis ad horizontem fuerit parallela, & quando parabolæ initium b, factum fuerit ex vertice supremo ipsius parabolæ, siue (quod idem est) ab extremo axis parabolici puncto b. Quando vero linea proiectionis a b non horizontalis, sed

33 Evangelista Torricelli (*De motu gravium naturaliter descendentium et projectorum, libri duo* (Florence, 1644) calls the study of projectiles 'the supreme fruit of Galileo's labours and his supreme glory'.

serious mathematicians expect their readers to have at least Euclid's *Elements* at their fingertips. The message is clear: without mathematics, no one can study motion. Aristotelians will simply disqualify themselves from studying physics if they neglect mathematics; in front of the book of nature they will be illiterate (8: 269–73).

Simplicio can still perform a useful function: he can make shrewd comments which enable Salviati to describe more nearly just what his new science of motion can achieve. Simplicio notices that if we represent the horizontal plane by a straight line then all its parts are not equidistant from the centre, so a body moving from the middle of the line will be going uphill; hence its motion cannot be perpetual nor uniform, but will always diminish until it stops. In any case, the resistance of the medium must destroy the uniformity of the horizontal motion and alter the rule of acceleration in fall. So it is very unlikely that things in reality will conform to demonstrations based on Salviati's suppositions.

Salviati concedes all this quite equably. His case is no different from that of Archimedes, who took the beam of a balance to be a straight line and treated threads hanging from its extremities as parallel. (It is very striking that Galileo had used this appeal to the practice of Archimedes in his first writings on motion (1: 300). So the one thing we are not seeing here is a novel departure from whatever there was of Aristotelianism in Galileo's early understanding of science.) In most cases, as it happens, the correction needed to make the demonstrations fit what actually happens wil be slight or negligible. The resistance of a medium such as air is, however, trickier. It will, for instance, impose a terminal speed on bodies which, as far as the rule of fall goes, would accelerate for ever. It will even interfere with and put an end to uniform horizontal motion. The scientific approach, says Salviati, is to disregard such accidental features and demonstrate the conclusions for the case of no impediments: the conclusions can then be used with such limitations as experience teaches. The limitations, Salviati points out, are not so great as Simplicio might suppose. Again he refers to falling bodies to make his point: in a fall of 150 or 200 cubits a wooden ball will reach the ground only a cubit or two behind a lead ball ten times as heavy, so neither ball is much affected by the resistance of the air. Hence there need be no great worry about disregarding the resistance of the medium in the demonstrations about projectile motion, since the weight of the balls makes little difference. Similarly, the resistance of the air is not much greater for rapidly moving bodies than it is for slower ones, as can be seen from the oscillations of pendulums of equal length with lead bobs of equal weight swinging through unequal arcs, one through 80 degrees from the perpendicular, the other through 5. So Salviati concludes that for the speeds he will be dealing with through distances which are short in comparison with the Earth's radius no important differences from what his demonstrations predict will be observed. Even the case of firearms presents no great practical problem, despite the 'supernatural fury' with which they can project bullets: the path of a high-velocity projectile might indeed be a somewhat flattened parabola, but the usual shots from mortars, which use only small charges, follow the parabolic path pretty nearly, so there is no difficulty in drawing up a table of ranges for given elevations of the mortar (8: 274–9, 304).

Salviati then launches into a series of theorems which take up the rest of the work. Simplicio is virtually silent until the concluding stages. Sagredo is able to keep up with this mathematical exposition of the new science of projectile motion, so he can insert a characteristic piece of praise both of mathematics and of Galileo's achievements:

> The force of necessary demonstrations which are found only in mathematics is at once full of wonder and delight. I already knew, on the authority of what most gunners say, that of all the shots of cannon or mortars the greatest, that is, the one that fires the ball the greatest distance, is the one made with an elevation of half a right angle, which they call the sixth point of the quadrant. But to understand the reason why this is so surpasses by an infinite interval the mere information derived from others' testimony or even from many repeated experiences.

Salviati naturally concurs:

> Your reasoning is very sound. Knowledge of a single effect acquired through its causes opens the mind to understand and ascertain other effects without having to resort to experiences, the present case being a good instance, where our Author, having ascertained by conclusive demonstration that the greatest of all the ranges was that when the elevation is half a right angle, shows what perhaps has not been observed by experience, namely, that of other shots those at elevations exceeding or falling short of half a right angle by the same amount are equal to each other, so that balls fired, one with an elevation of seven points and the other with five, will strike the horizontal at the same distance. (8: 296)

So shipwrights, builders, hydraulic engineers and now gunners have all been catered for in the *Discourses*. At the same time, the strength of the work, with all its simplifications and idealizations, is a mathematical reading of important parts of the book of nature. The actual mathematics employed by Galileo, the language of proportions from Euclid, was soon to be replaced by more adequate tools. His followers, unlike contemporary followers of Aristotle, were not only ready but eager to correct their teacher's findings. They had, however, no thought of disowning his general programme: the 'Archimedean' approach he had embarked on five decades earlier proved to be a decisive innovation. Its decisiveness is hardly lessened even if we grant that he owed much to his medieval predecessors. One does have to mention that in a short biography it is not possible to do justice to them or to his immediate predecessors and contemporaries. This point is of special importance in the case of Galileo, who had a weakness for crying up his own originality. But he was far more than an arranger of unacknowledged borrowings. Nor was his decisiveness a mere matter of assuming the leadership of the 'novelists' and becoming the most persuasive advocate of new discoveries and new methods. In astronomy he was the originator in making telescopic discoveries of significance; his leadership in promoting those discoveries and turning them to Copernican account was made possible by his own acknowledged achievements. In physics the new science of motion, with its wedding of experiment and mathematics, has to count as his. In his own way he was even a decisive innovator in theology, though there his efforts came to nothing. If he laid waste significant tracts of the Aristotelian empire, he had something to put in its place and he could tell himself with some plausibility that he had done all this methodically in a way that Aristotle would find hard to disown.

Last Years

The Florentine Inquisitor was correct in thinking that the aged Galileo would never again attempt to promote Copernicanism. In fact, within a year of his death Galileo wrote in a letter that the falsity of Copernicanism must not on any account be called into doubt, especially by Catholics. All Copernican conjectures to the contrary are removed by the most solid argument from God's omnipotence. The supposed solidity of the argument so dear to Urban VIII did not preclude Galileo from pointing out that Ptolemaic and Aristotelian arguments were not only inconclusive but even more fallacious and erroneous, as long as one remained within the limits of human reasoning. It is evident both that Galileo had not changed his mind and that he had resigned himself to the fact that his own part in the campaign to establish Copernicanism was over (18: 314–15). Still the Roman Inquisition, as a matter of course, ruled out any discussion of Copernicanism when granting Castelli's petition to visit Galileo for a twofold purpose: to assist him in preparing spiritually for death and to discuss the method of finding longitude at sea (17: 406). Yet Galileo cannot always have been under very strict surveillance. A visit by John Milton to Arcetri, probably in the autumn of 1638, does not seem to have ruffled anybody.[8]

The Inquisitor was also correct in his general assessment of Galileo's failing health. Yet, in spite of everything, Galileo did remain remarkably active intellectually. For this a good deal of the credit must go to Viviani, who came to live at Arcetri as Galileo's pupil in October 1639.

With the help of Viviani and others Galileo carried on an important correspondence with a very capable Genoese, Giovanni Battista Baliani, whom he had known for nearly a quarter of a century. Baliani was currently working independently on many of the central topics which Galileo had treated in the *Discourses*, so he was able to draw out Galileo on whether the demonstrations of the two new sciences showed how things really are. Galileo again likened his method to that of Archimedes, this time referring to Archimedes's investigation of spirals. The fact that such spiral motions of bodies may not be found naturally tells nothing against the proofs adduced. Similarly, if in nature we did not find anything like bodies falling in accord with the mathematical definitions and demonstrations in the *Discourses*, that would not matter much to Galileo. But he had been lucky: the way bodies fall fits in exactly with motion as he has defined it (18: 12–13). It should be noted that this is perfectly compatible with the realism (as distinct from instrumentalism) which characterized Galileo's campaign for Copernicanism. For strategic reasons he had in the *Discourses* postponed the search for the causes of motion. There is no clear evidence that he had abandoned the search for a true understanding of the real world, of the real, unique constitution of the universe, including the motions within it. It is also interesting to notice Baliani reporting in September 1639 his own test of whether a stone dropped from the mast of a moving ship falls at its foot. That it did was, of course, no surprise to Galileo (18: 103).

Another fascinating correspondence was with Fortunio Liceti, who had begun to teach philosophy at Padua in Galileo's last academic year there. Liceti was a prolific author of Aristotelian works. What Galileo clearly found attractive in this not very distinguished thinker – the very type of philosopher whom Galileo had demolished more times than he cared to remember – was the unaffected and unfailing courtesy with which Liceti approached him as a fellow-student of the truth. Thus in September 1640 Liceti was very pleased to hear that Galileo did not consider himself opposed to Aristotle, though he freely confessed that this had come as a surprise, since he had had the contrary impression from Galileo's writings; still, like many others who had had this impression, he could be mistaken in this particular (18: 245). Such a courteous prompting to explain himself more fully was not wasted on Galileo, who immediately dictated a lengthy reply.

To be an Aristotelian philosopher consists chiefly in following what Aristotle taught about scientific reasoning, especially about how to avoid fallacies and deduce conclusions correctly. Galileo claims to have learned from mathematics how to reason safely: in fact, any fallacious reasoning in his works will be very rare indeed. So to that extent he is an Aristotelian. Next, he agrees with Aristotle in placing experience above any argument or appeal to authority, *even one's own*. Now he will judged to be opposed to Aristotle by all those who think that to philosophize well is to defend whatever Aristotle said, even if this means denying sense experience or making Aristotle say things that never entered his head. There is nothing odd in the fact that Aristotle should have discovered the correct principles of method and yet himself have made mistakes in applying them, just as a painter may inadvertently make a mistake in perspective. Should Galileo ignore such mistakes out of deference to Aristotle's authority? No. If Aristotle returned to the world, he would much rather count as a follower Galileo, with his few but well-founded rejections of Aristotelian ideas, than the many who try to defend texts by imposing on them meanings Aristotle never thought of. If Aristotle could see the novelties recently discovered in the heavens, he would certainly change his opinion on the immutability of the heavenly bodies (18: 247–50).

Nor did Galileo limit himself to reflections on his own published work. Although he was acutely conscious of how reduced his intellectual powers were he continued to work on new ideas and to follow those of others. He was particularly grateful to Castelli, his most faithful pupil (18: 179). Cavalieri, too, kept in touch, though his hope of the 'triumvirate' meeting at Arcetri was never fulfilled, partly because he and Castelli were almost as plagued by poor health as was Galileo himself. Castelli did manage to visit Galileo in March 1641, bringing as an extra consolation a book by his pupil, Evangelista Torricelli (1608–47), a work which would let Galileo see that the way he had opened to the study of motion was being followed by able minds (18: 303). Torricelli actually came to live at Arcetri in October 1641, to Galileo's great delight (18: 368), but they had little time to work together because by mid-November it was evident that Galileo was dying. He died during the night of 8 January 1642 and was buried privately in Santa Croce.

A public funeral with great solemnity would normally have followed and the Grand Duke wished to erect a suitable commemorative mausoleum. Cardinal Francesco Barberini, writing on behalf of the Pope and the Inquisition, made it plain that such recognition was not appropriate for one who, though he had died a good Catholic, was nevertheless still doing penance for a serious offence. If there had to be a monument with an inscription it must avoid anything that could harm the reputation of the Inquisition. The same applied to any funeral oration. The Pope spoke in similar terms to the Florentine ambassador.

He and his nephew were concerned to avoid giving scandal to the good. The Grand Duke, on the advice of the ambassador, bowed to their pressure (18: 378–80).

Urban VIII and his nephew were merely acting consistently with what they had done before. The opportunity for magnanimity had already been missed in Galileo's last years. That *their* actions might offend good people both at the time and later was probably beyond their imagining. In any case, Galileo needed no monument except his own writings. He had achieved his ambition to obtain some splendour from his studies.

10
Rehabilitation

In the autumn of 1992 people learned from television and newspapers that Galileo had been rehabilitated by Pope John Paul II. The brief but accurate newspaper reports were sometimes understandably accompanied by simulated incredulity and speculations about what else the Catholic Church might catch up with in another three and a half centuries. But it was generally understood that it was not Galileo who needed to be rehabilitated. The whole point of so-called rehabilitations, whether political or ecclesiastical, is to acknowledge wrong caused by predecessors in government of State or Church to a member or members of the State or Church concerned. In a way it is the relevant authority that is being rehabilitated. It is rather like the Catholic understanding of Confession: the past cannot be changed, but it is crucial to say sorry and make a firm purpose of amendment for the future. It should also be noted that the word 'rehabilitation' was an editorial contribution, but it was an apt enough interpretation of what the Pope intended.

On this occasion no one saw the event as marking a novel departure in Catholic theology. As we noticed in chapter 8, most Catholics who took any interest in the matter were free to adopt Newtonianism once the *Index* of 1758 omitted the general ban on books advocating the motion of the Earth. It is true that in 1820 an obstructive official in Rome used legalistic arguments to prevent publication of an astronomical work, but he was soon overruled and the consequence was that Galileo's *Dialogue* was duly removed from the next edition of the *Index*.

34   Baden Powell's *Essays on the Spirit of Inductive Philosophy*: the device
on the back cover of this edition (London, 1855) shows the papacy keeping
the earth fixed centrally.

For the next century or so, the aftermath of Galileo's condemnation is most noticeable for its effects on Catholic apologetics: the general aim was damage limitation in the face of the sweeping charge that the Galileo case was only the most signal instance of Catholic obscurantism and hostility to science and secular learning in general. The contribution of Catholic apologists played its own useful part in building up a much more critical and solidly documented history of science, so that (at least in the academic world) sweeping theses about the enduring and inevitable conflict between science and religion now serve principally to alert undergraduates to the dangers of myth-eaten slogans posing as documented conclusions. It is not too optimistic to say that for some decades now the Galileo case has provided no excuse for avoiding undoubtedly genuine problems in sociology of science, philosophy of science, philosophy of religion and theology. Indeed, even a short study of Galileo's life and his condemnation still raises many such problems. In that sense the Galileo case may be, as the sub-title of

Fölsing's biography suggests, *Prozess ohne Ende*, an endless trial or case. But many Catholics have long felt that the condemnation of Galileo was for the Church a very specific piece of unfinished business and one that could be brought to completion pretty straightforwardly and speedily.

The Second Vatican Council

The Second Vatican Council, a formal gathering of the bishops of the Roman Catholic Church, met for about two months each autumn of the years 1962 to 1965. One of the most important documents it produced was the pastoral constitution on *The Church in the World of Today*. Since the principal concern of that document was to address all people of goodwill in a spirit of dialogue and cooperation, it was essential that it should say something about the relation of faith to science, especially since in common perception, and even in actual fact, the Catholic Church (or at least many of its members) still seemed suspicious or grudging in its attitude to the achievements of scientists. A Church which was convinced that there is no endemic warfare between religion and science could afford to implement that conviction in a more appreciative recognition of the proper autonomy of the sciences and of human intellectual and artistic pursuits in general.

In this context it was to be expected that the Galileo case would receive some attention. In fact, several bishops referred to it in speeches or written proposals. Most of them saw it as a warning to the Church not to trespass beyond its remit, though it is interesting to notice that two bishops thought that too much fuss was being made of the Galileo affair. One of the most striking interventions both on the lingering ecclesiastical suspicion of secular culture and on the Galileo case itself came from Bishop Elchinger, coadjutor of Strasbourg, in a speech to the Council on 4 November 1964.

Many people, he said, still took the Galileo case as typifying the various shortcomings of the Church in its appraisal of human culture. It was not just a matter of past history:

> In this, the fourth centenary of the birth of this worthy man, many scientists throughout the world are celebrating his memory, but right up to today no reparation has been made for that wretched, unjust condemnation. In the world of today *acts* are more important than words. The rehabilitation of Galileo carried out by the Church, humbly and properly, would be an eloquent action.[1]

Elchinger did not succeed in securing the rehabilitation he requested, perhaps because it was not the kind of work which could be adequately undertaken in the time available by the hard-pressed editors of the document. But he, and those bishops who made similar interventions, were rewarded by a discreet reference in a footnote to a biography of Galileo. This 'clear allusion', as it was officially described in the report to the bishops in Council, made it plain that Galileo was a victim of what the text of the document referred to, namely certain deplorable attitudes, sometimes found even among Christians, which derive from insufficient regard for the legitimate autonomy of science. The consequent conflicts and controversies have led a good number of people to think that faith and science are opposed to each other.[2]

A Speech on Einstein's Centenary

Pope John Paul took up the Galileo case again in a speech he delivered in French on 10 November 1979, on the occasion of the centenary of Einstein's birth. His audience was the Pontifical Academy of Sciences (which, as the Pope noted in passing, counts itself as a successor of the Lyncean Academy), and the College of Cardinals. His general theme was how the Church admires and welcomes the achievements of scientists. I shall pass over his tribute to Einstein and pick out only what he said about Galileo: his greatness, like Einstein's, is known to all.

But we cannot hide the fact, says the Pope, that Galileo had to suffer much from ecclesiastical men and institutions. He then quotes the passage from the Council which alluded so clearly to Galileo as a victim of deplorable attitudes. Then comes perhaps the most important passage:

> To move beyond this position taken up by the Council, I should like theologians, scientists (savants) and historians, inspired by a spirit of sincere collaboration, to go further into the Galileo case, and by impartial recognition of mistakes from whatever source, to remove from many minds the blockage which that affair still puts in the way of a fruitful concord between science and faith, Church and world. I give full support to this task, which will allow the truth of faith and of science to be respected and will open the door to further work together.

We shall see in a moment the practical support which the Pope had in mind, but first it is worth noticing some further remarks which he made. Galileo, he says, who is rightly called the founder of modern physics, declared explicitly that the truths of faith and science cannot contradict each other. In developing this point the Second Vatican

Council uses expressions similar to Galileo's in his *Letter to Castelli*. The Pope recalls Galileo's genuine piety (which I mentioned in chapter 1) expressed in his *Sidereal Message* and acknowledges (in the passage already quoted at the end of chapter 6) that, in his *Letter to the Grand Duchess*, Galileo formulated important epistemological norms which are indispensable in reconciling scripture and science.

The Pope recognizes that not all questions arising from the Galileo case are solved by an acceptance of the general principles he indicates in his speech. But the principles (clearly laid down in the Second Vatican Council) concerning the legitimate autonomy of science and religion do help to provide a favourable starting-point for an honest and impartial dismantling of old oppositions. This whole speech was, if not technically a rehabilitation of Galileo, certainly a generous acknowledgement not only of the irreplaceable contribution of the sciences to any fully human flourishing, but also a tribute to Galileo's own courageous attempt to integrate the resources of both science and faith.[3]

The Dialogue *Revisited*

The practical follow-up to the Einstein centenary speech was the establishment on 3 July 1981 of an interdisciplinary Pontifical Commission to investigate the Galileo case further. During the following ten years members of the Commission produced a series of impressive publications on various aspects of the context and the implications of the Galileo affair. While it was at work the Pope more than once emphasized the importance of cooperation between scientists and theologians for the benefit of humankind. He also used the occasion of a symposium, held to mark the three hundred and fiftieth anniversary of the publication of the *Dialogue*, to reiterate his support for the work of the Commission. His audience on 9 May 1983 was composed of scientists from many countries, the Cardinals resident in Rome, and other Church dignitaries.[4]

From the Galileo affair the Church has learned 'a more accurate understanding of its own authority'. At the Second Vatican Council the Church expressed its support for the freedom of research, confident that there can be no contradiction between science and faith. They are, says the Pope, distinct, autonomous orders of knowledge, which finally converge on the discovery of that integral reality which has its origin in God.

So it becomes clearer that divine revelation, of which the Church is the guarantor and witness, does not of itself imply any scientific theory of the universe, and the assistance of the Holy Spirit in no way goes surety for the explanations of the physical constitution of reality which we may wish to maintain.

Although the Pope does not say so, that statement is certainly much closer to Galileo's views on scripture than to Bellarmine's (let alone Caccini's). But the Pope reasonably points out that we should not be surprised if such a complex matter caused the Church great difficulty, and he explicitly reminds his audience that some, notably Bellarmine, did wish to avoid useless tensions and harmful rigidities in the relations of science and religion. Here the Pope is quoting from a speech he made to Bellarmine's own Gregorian University on 15 December 1979. The rest of the speech leaves no doubt that the Pope's interest in the Galileo case is just one important part of a comprehensive attempt to bring the two cultures of science and religion into useful cooperation based on mutual respect.[5]

The Commission's Conclusions

On 31 October 1992 Cardinal Paul Poupard presented the Pontifical Commission's conclusions in summary form to the Pope in the Sala Regia of the Apostolic Palace. The Cardinal first explained that it was not a question of conducting a retrial. He did not elaborate on this point but it may be suggested that a retrial is a very dubious notion. Unless it could be shown that the procedures in force at the time were not carried out properly, what would it mean to say that the condemnation of Galileo was somehow void? That Church officials should have known better? That the legal procedures in the Church at that time gave inadequate protection to the accused? Or that we now see that Galileo was more in the right than his accusers? It is quite possible that if a genuine retrial were staged (supposing there could be such a thing), it would lead to a similar result, though possibly with a lighter sentence, or perhaps nothing worse than a 'correction' of the text of the *Dialogue*. (This brings out the artificiality of a retrial after three and a half centuries.) Could anything that counted as a retrial of Galileo get round the fact that the official intervention of 1616 could not be ignored and that Galileo's *Dialogue* would need extremely benign interpretation if it were to be seen as compatible with that intervention? Could there be a 'retrial' of the intervention of 1616, which is where it all went wrong, when that intervention could not be called a trial in any clear sense? (It is unfortunate that the Cardinal refers once to 'the two trials'.) The whole notion of a retrial is suspect, if it amounts to anything more than a decent wish that the whole episode had never happened or at least not been so needlessly humiliating to a great man. The Commission was surely right in its judgement that interdisciplinary work offered a much more straightforward way of assessing the rights and wrongs of the affair. Their brief, as Cardinal Poupard says, was to answer the questions: What happened? How did it happen? Why did it happen?

The Cardinal first expounds Bellarmine's insistence that real proof of Copernicanism should be brought before there could be any thought of reinterpreting scripture. Galileo could not produce irrefutable proofs, but neither could his adversaries, then or later, discover a convincing refutation of Copernicanism. The summary of Bellarmine's position, as presented in the *Osservatore Romano*, includes a sentence which could mislead a reader:

> According to Robert Bellarmine, as long as there was no proof that the earth orbited round the sun, it was necessary *to interpret with great circumspection* the biblical passages declaring the earth to be immobile.[6]

It is true that Bellarmine, who correctly thought that Copernicanism was unproven, advised Galileo and Foscarini to be cautious. It is also true that he said that the reinterpretation of scripture should not be ruled out hastily. All this was known to Galileo, as we saw in chapter 6. But not only did Bellarmine move very smartly to execute the (not very circumspect) orders of Paul V. He used the phrase 'with great circumspection' to describe how the Church should proceed if a conclusive proof were produced. In his next sentence Cardinal Poupard correctly describes Bellarmine's allowing that if there were a proof of Copernicanism, then theologians would have to review their interpretation of biblical passages whose meaning was in dispute. Only if such a proof were produced would Bellarmine consider reinterpretation; it is in those circumstances (even in those circumstances) that 'great circumspection' would be called for. This part of the summary would have been a clearer reflection of the scholarly standards of the Commission's publications if it had stayed closer to Bellarmine's words throughout (they are correctly quoted in the original Italian in a footnote). It remains true, of course, that as long as there was no proof of Copernicanism, Bellarmine required circumspection *and more* from Galileo and Foscarini.

Cardinal Poupard then briefly relates how the general prohibition of

Copernican books disappeared from the *Index* and how the last traces of official anti-Copernicanism were removed in 1822. The Cardinal merely describes this history. The Commission does not seem to have tackled thoroughly the question of why it had taken so long – at least until 1757 – for the original mistake to be recognized. Even granted that what could count as conclusive optical and physical evidence of the falsity of geocentrism was not available until the detection of annual stellar parallax or the demonstration of Foucault's pendulum (the Cardinal seems to allude to these), it is not explained why the Church officially allowed heliocentrism to be taught as true before such evidence was available (unless it is claimed that it was the detection of the aberration of starlight which clinched the matter). Nor does the Cardinal's report offer any helpful reflections on the position of theologians in the century after Galileo's condemnation.

Given that the Inquisition and the Congregation of the Index had intervened in 1616 to rule that Copernicanism could be held only as a calculating device and given the implementation of that ruling in the condemnation of Galileo in 1633, where did that leave theologians? They knew of course that the decisions taken were reformable; they also knew that technically the condemnation of Galileo was a purely personal matter, but in practice it was implemented as a warning to all Catholics. On any reading of the history it took a long time for any official relaxation of the ban on heliocentrism to be quietly lifted – much more quietly than it had been imposed. If there is a legitimate autonomy for science, should there not also have been a legitimate autonomy for theology, at least enough to allow theologians, as part of their professional task, to help to correct as soon as possible what later had to be admitted on all hands to have been a mistake? Why did one have to be something of a maverick in the profession to argue openly, during the century after Galileo's condemnation, that the Church ought to acknowledge its mistake? Did theologians just have to accept that they were under an authority which was swift to anger and slow to relent? Can mistakes be corrected only when they no longer matter to anyone, when they have the status of fossils? There seems to me to be some further unfinished business arising out of the Galileo case that could be usefully addressed.

The Commission concludes that all those involved in the trial, without exception, have a right to the benefit of good faith, in the absence of extra-procedural documents to the contrary. Astronomy was in transition and scripture scholars were confused about cosmology, so the official scientific and theological assessment of Copernicanism was mistaken (this presumably refers to 1616). Certain theologians failed to grasp the profound, non-literal meaning of the scriptures when they describe the physical structure of the created universe. This led them unduly to transpose a question of factual observation into the realm of faith.

Again, this seems fair and welcome comment. The Commission does seem to have seen its remit as at least implying that certain issues need to be addressed if similar mistakes are not to recur. But whatever the Commission's terms of reference were, the Galileo case naturally raises questions not just about the judicial procedures of that case, but also about whether anything could be learned to help in future difficulties. What the Commission (and the Pope himself; see below) say about ways of avoiding needless conflict in the future is clear and helpful. Nevertheless one cannot avoid questions about current *procedures*: would they be adequate if, despite the best efforts of many people, some such apparent conflict as faced Galileo and Bellarmine were to occur now? Naturally the procedures in use in the Catholic Church nowadays (as in civil law) are very different from those obtaining in Galileo's time, when torture or even burning was a real threat. Even Bellarmine's perceptiveness and courtesy cannot disguise the fact that the actual consultation on which the 1616 intervention was based was routine and casual. Advice was given by consultors who were not only mistaken theologically, but were not qualified to understand the scientific issues, so it is not surprising that Caccini and Lorini carried more weight than Galileo.

Now the successor to the Inquisition, the Congregation of the Doctrine of the Faith, does not have a merely negative role: part of its brief is positive; it is expected to make suggestions and proposals and to offer counsel. At the same time it is still entrusted with the difficult task of assisting the Pope in the delicate work of guarding orthodoxy, so it has to be ready to conduct investigations of writings it considers suspect. Even in such investigations, which are always liable to be the occasion of controversy, the Congregation tries to shape its exchanges with the author on the model of a critical dialogue rather than on the sort of high-handed judgement which, as Urban VIII told the Florentine Ambassador, was the way things were done by the Inquisition. (It is tempting to imagine how such a critical dialogue might have gone in 1633.) But the genuine intention to carry out any investigation in the spirit of dialogue does not lessen the need for adequate legal procedures to protect the rights of the person under investigation. The complaint is still made frequently by theologians and Church lawyers that the judicial procedures employed by the Congregation compare very poorly with those which have long been taken for granted by the citizens of many countries. Indeed, the Congregation itself seems to admit that its procedures are in need of improvement, so there is still work to be done.[7] It seems that the work of devising better procedures has been postponed for nearly a decade because the very small staff of the Congregation has been too overworked to attend to it. But one thing the Church is not short of is lawyers competent enough to draw up respectable procedures, so it is hard to see why there should be such delay. If this is new business, rather than something which the Commission on the Galileo affair might have been expected to comment on, perhaps there is a case for establishing another Pontifical Commission to tackle it.

The main conclusion of the Commission is printed in italics in the English translation of the Cardinal's speech:

> It is in that historical and cultural framework, far removed from our own times, that Galileo's judges, incapable of dissociating faith from an age-old cosmology, believed, quite wrongly, that the adoption of the Copernican revolution, in fact not yet definitely proven, was such as to undermine Catholic tradition and that it was their duty to forbid its being taught. This subjective error of judgement, so clear to us today, led them to a disciplinary measure from which Galileo 'had much to suffer'. These mistakes must be frankly recognized, as you, Holy Father, have requested.

It is clear from the appended list of the Commission's publications that the Cardinal could have given a much fuller presentation of the members' work and conclusions. But on this occasion the important speech, though no doubt written at least in part by the Commission, had to come from the Pope and there would have been little point in having everything said twice or in putting all the interesting bits in the Cardinal's summary presentation of the Commission's findings.

Pope John Paul's Speech

In his speech of reply to the members of the Pontifical Academy and other distinguished guests, the Pope thanked Cardinal Poupard and all who had taken part in the work of the Commission. He thanked them especially for their scholarly publications, which he valued highly. In future it would be impossible to ignore the Commission's conclusions. But why did he wish to return to the Galileo case? 'Has not this case long been shelved and have not the errors committed been recognized?'[8] True though that is, says the Pope, one cannot exclude the possibility that a similar situation could arise, one which would require both sides (science and faith) 'to have an informed awareness of the field and of the limits of their own competencies'.

What about the debate which centred on Galileo? Well, continues the Pope, Galileo, like most of his adversaries, did not distinguish between the scientific approach to natural phenomena and the philosophical reflection usually called for by that approach. 'That is why he rejected the suggestion made to him to present the Copernican system as a hypothesis, inasmuch as it had not been confirmed by irrefutable proof.'

What I have suggested in earlier chapters about Galileo's over-confidence in the case for Copernicanism is less tidy than the kindly but summary explanation which the Pope gives in this passage. I would rather say that both Galileo and his adversaries habitually devoted time to reflection which we would count as philosophical, but they still remained at cross-purposes and neither party could solve all the philosophical issues involved in their dispute. I also think that even a very brief summary of the episode, if it is not to be misleading, has to touch on the ambiguity of the crucial word 'hypothesis'. One could certainly say that neither Galileo nor Bellarmine was completely successful in sorting out the status of hypotheses, but there were solid reasons why Galileo could not accept the advice offered him (by Bellarmine). Pope John Paul moves on to theologians:

> The problem posed by theologians of that age was, therefore, that of the compatibility between heliocentrism and scripture.
>
> Thus the new science, with its methods and the freedom of research which they implied, obliged theologians to examine their own criteria of scriptural interpretation. Most of them did not know how to do so.

That is crisp enough for anyone (though theologians could defend their predecessors by saying there wasn't much room left for examining their criteria once the official ruling of 1616 had been made). The Pope caps this with a generous tribute to Galileo the theologian: 'Paradoxically, Galileo, a sincere believer, showed himself to be more perceptive in this regard than the theologians who opposed him.'

On this particular occasion it would not have been appropriate for the Pope to mention the incidental shortcomings of Galileo's *Letter to the Grand Duchess*. Instead, with justified magnanimity, he says it is like a short treatise on biblical hermeneutics. By implication we can see that Galileo did engage in the 'epistemological reflection on the biblical sciences' demanded by the Copernican upset, whereas theologians contemporary with him did not, or at least did not do so with any very helpful results, so that the 'abundant fruit in modern exegetical works' came only much later.

Naturally the Pope recognizes that the

> pastoral judgement which the Copernican theory required was difficult to make, in so far as geocentrism seemed be a part of scriptural teaching itself . . . Let us say, in a general way, that the pastor ought to show a genuine boldness, avoiding the double trap of a hesitant attitude and of hasty judgement, both of which can cause considerable harm.

The Pope, we may presume, is well aware that such a generality is only a limited help, given that it was more or less the position which Bellarmine took, and conveyed to Galileo through Dini, as we saw in chapter 6. Generalities of this kind undoubtedly have their uses, if only to remind us that anyone can be too hasty or too hesitant and that there is no recipe for judiciousness that can be applied without hard intellectual work. To advocate 'genuine boldness' does nothing to identify what is the genuinely bold decision in particular circumstances, but it *is* a useful reminder that there is such a virtue, even if it is liable to be classed as hastiness by opponents. In this instance some bite is given to the generality when the Pope gives an example of a 'hasty and unhappy decision'. Referring to a new understanding of the Bible and the biblical world made possible by advances in the historical sciences at the end of the last century and the beginning of this, he says that certain people, in their concern to defend the faith, thought it necessary to reject firmly based historical conclusions. He also pays a tribute, which is handsome enough even in the lumpy English of the translation: 'The work of a pioneer like Fr Lagrange [a Dominican scripture scholar] was able to make the necessary discernment on the basis of dependable criteria.'

There is no mention of the fact that the 'certain people' included the Pontifical Biblical Commission and Pius X. That is the usual style of such documents. The frequent papal custom of quoting predecessors is considered inappropriate except to establish continuous development, which is why Paul V and Urban VIII are not mentioned in connection with Galileo. But the positive thrust of the speech is clear:

> It is a duty for theologians to keep themselves regularly informed of scientific advances in order to examine, if such be necessary, whether or

not there are reasons for taking them into account in their reflection or for introducing changes in their teaching.

This point is reinforced by quotation (admittedly selective) from Bellarmine and Augustine: a reader could be left wondering how Bellarmine did not manage to get it right. It may be that the Pope is merely extending to Bellarmine the sort of benign interpretation he gave to Galileo's *Letter to the Grand Duchess*.

One would not expect the Pope, or even the Commission, to give a survey of the work of theologians, philosophers and scientists in the three and a half centuries since Galileo's death. But the Pope's mention of 'boldness' is more than an empty generality. In this short sketch of Galileo's life and works I have included a fair amount of material on the work of Jesuits contemporary with Galileo. The Society of Jesus is rightly credited with many achievements which only boldness could have secured, but at the time that mattered to Galileo their official policy in philosophy (including science) and theology was avowedly and deliberately cautious. I have tried to sketch that cautious policy sympathetically, since it is not hard to sympathize with a Grienberger, a Guldin or a Grassi. But it is harder to respect the decision to reinforce that policy in the eighteenth century. There must come a stage when unremitting caution brings greater damage to theology and philosophy than does the occasional bold sally, which may or may not be successful. At all events, Pope John Paul does envisage that scientific advances may give theologians reasons for introducing changes in their teaching, so it cannot be excluded that such changes would help to reshape the official teaching of those in authority in the Church.

Pope John Paul next describes the way that since the Enlightenment the Galileo case has been made into a sort of 'myth', the myth of dogmatic obscurantism opposed to the free search for truth. '*A tragic mutual incomprehension* has been interpreted as the reflection of a fundamental opposition between science and faith.' This sad misunderstanding now belongs to the past. Galileo comes in for handsome praise as 'a brilliant physicist', who 'practically invented the experimental method'. The Pope's laconic remark that in fact the Bible does not concern itself with the details of the physical world is the voice of Galileo, of which the following passage is also an echo:

There exist two realms of knowledge, one which has its source in Revelation and one which reason can discover by its own power. To the latter belong especially the experimental sciences and philosophy. The distinction between the two realms of knowledge ought not to be understood as opposition. The two realms are not altogether foreign to each other; they have points of contact.

The remainder of the speech is taken up with outlining how the Pontifical Academy promotes the advancement of knowledge, with respect for the legitimate freedom of science.

> What is important in scientific or philosophic theory is above all that it should be true or, at least, seriously and solidly grounded. And *the purpose of your Academy* is precisely *to discern and to make known*, in the present state of science and within its proper limits, *what can be regarded as an acquired truth* or at least as enjoying such a degree of probability that it would be imprudent and unreasonable to reject it. In this way unnecessary conflicts can be avoided.

After further wide-ranging reflections on the relations of science and faith the Pope leads into his concluding thanks with this:

> [The] intelligibility, attested to by the marvellous discoveries of science and technology, leads us, in the last analysis, to that transcendent and primordial Thought imprinted on all things.

This is an eloquent conclusion, which Galileo would have applauded, to the 'eloquent action' which Bishop Elchinger had asked for. The 'rehabilitation' is ungrudging in its acknowledgement of what Galileo suffered and what he achieved, both as a scientist and a theologian. However belated, it is a posthumous Roman triumph for the decisive innovator.

Notes

CHAPTER 1   *THE STRANGEST PIECE OF NEWS*

References to the *Edizione Nazionale* in the notes are by volume and page number, as in the text, preceded where necessary by the abbreviation *EN*.

1   There is a very useful English translation, with introduction, conclusion and notes by Albert Van Helden: *Sidereus Nuncius or The Sidereal Messenger*, University of Chicago Press, 1989.
2   I have used the transcript made by Logan Pearsall Smith, *The Life and Letters of Sir Henry Wotton*, 1907, vol. I, pp. 486–7. A fuller excerpt is given, without reference, by I. Bernard Cohen, *The Birth of a New Physics*, 2nd edn, 1985, pp. 75–6.
3   John W. Shirley, *Thomas Harriot*, 1983, p. 397.
4   Ibid., p. 399.
5   Stillman Drake, *Galileo: Pioneer Scientist*, 1990, p. 6.
6   Ptolemy, *Almagest*, Book I, ch 7; translated and annotated by G. J. Toomer, 1984, p. 45.
7   A. Van Helden, *Sidereus Nuncius*, p. vii.
8   On the scientific revolution see I. Bernard Cohen, *The Newtonian Revolution*, 1983, and David C. Lindberg and Robert S. Westman (eds), *Reappraisals of the Scientific Revolution*, 1990.
9   Kepler's *New Astronomy* of 1609 was, as the title claimed, *aitiologetos*, a reasoning out of causes or a celestial physics.
10   See Victor E. Thoren, *The Lord of Uraniborg: A Biography of Tycho Brahe*, 1990.
11   See the excellent article by Mario Biagioli, 'Galileo's System of Patronage', *History of Science* 28 (1990), pp. 1–62.

12 A. Van Helden, *Sidereus Nuncius*, pp. 3–4. See also his 'The Historical Problem of the Invention of the Telescope', *History of Science* 13 (1975), pp. 251–63.

13 Mary G. Winkler and Albert Van Helden, 'Representing the Heavens: Galileo and Visual Astronomy', *Isis* 83 (1992), pp. 199–217, especially p. 215. I am very grateful to Dr Mary T. Brück for giving me a copy of the illustrated paper, 'Galileo's Telescopes and his Astronomical Discoveries', written jointly with her husband Professor Hermann Brück, which she delivered at the Conference 'Galileo and his Times' held at the Royal Observatory, Edinburgh, on 18 November 1989.

14 For the dating of Galileo's lunar observations and a critical discussion of his published and unpublished drawings of the Moon (which show his artistic skill) see Ewan A. Whitaker, 'Galileo's Lunar Observations', *Journal for the History of Astronomy* 9 (1978), pp. 155–69 [with photographs].

15 Simon Mayr, often known by his Latin name of Marius, published his own observations in his *Mundus Iovialis* of 1614. Galileo, with some serious grounds for suspicion, treated him as a plagiarist. Some still do, but his reputation was vindicated earlier this century. See the entry on him by Edward Rosen in *Dictionary of Scientific Biography*, vol. 9, pp. 247–8.

16 See Edward Rosen's *Kepler's Conversation with Galileo's Sidereal Messenger*, first complete translation, with an introduction and notes, *The Sources of Science*, no. 5, Johnson Reprint Corporation, New York and London, 1965.

CHAPTER 2   *EARLY LIFE*

1 *EN* 19: nativity charts 23–4; baptism 25; Vincenzio Galileo's notes 594–6; Viviani's historical account 599–632; Gherardini's life 633–46; inscription 11. For the inscription I have used the appendix of 1702 to Book 3 of Viviani's *De locis solidis secunda divinatio geometrica*, pp. 126–7.

2 Antonio Favaro, *Galileo Galilei e lo studio di Padova*, 2 volumes first published in 1883 and reprinted in 1966, vol. I, p. 6.

3 The exercise is at 9: 282–3. A note by Galileo on the cover of a book (20: 586) says that on 10 August 1604 he began to translate into Italian verse the *Batrachomyomachia*, then attributed to Homer. He may have been translating from a Latin version. Viviani (19: 601) says his knowledge of Greek was 'not mediocre' and that he used it in his more serious studies.

4 Gherardini says that Vincenzio had at first thought of putting Galileo to the wool trade rather than studies: 19: 635.

5 The date written in the record is clearly 1581 but, because the calendars in use at Florence and Pisa can cause confusion about the year intended, Favaro noted the possibility that Galileo matriculated in 1580 rather than 1581: *Galileo Galilei e lo Studio di Padova*, vol. I, pp. 8–9. Schmitt refers to an article which I have not seen: this seems to settle on 1580. See Charles B. Schmitt, *Reappraisals in Renaissance Thought*, ed. Charles Webster, 1989, X, p. 27, referring to R. Del Gratta, 'A proposito della data d'iscrizione di

Galileo Galilei all'Università di Pisa [1580 settembre 5; e non 1581!]', *Bolletino storico pisano*, 46 (1977), pp. 556–8. For the University of Pisa I have relied heavily on Schmitt's articles collected in the volume cited and in his two earlier volumes published by Variorum Reprints, London: *Studies in Renaissance Philosophy and Science*, 1981; and *The Aristotelian Tradition and Renaissance Universities*, 1984.

6   He twice gives a classical authority on fol. 12 (E & H) of his *De motu libri X*, Florence, 1591.

7   See, for instance, David C. Lindberg (ed.), *Science in the Middle Ages*, 1978, and most recently his *The Beginnings of Western Science*, 1992. See also Allan Franklin, *The Principle of Inertia in the Middle Ages*, 1976.

8   Ptolemy himself, in other works, was no instrumentalist. Even his *Almagest* can be read as an attempt to describe the true system of the universe as nearly as possible; but in Galileo's time an intrumentalist interpretation was very influential. For a sympathetic survey of this tradition see Pierre Duhem, *To Save the Phenomena: An Essay on the Idea of Physical Theory from Plato to Galileo*, trans. Edmund Doland and Chaninah Maschler, 1969. For a realist interpretation of Ptolemaic astronomy see Peter Barker and Roger Ariew (eds), *Revolution and Continuity: Essays in the History and Philosophy of Early Modern Science*, Studies in Philosophy and the History of Philosophy, vol. 24, 1991, p. 5 [with further references].

9   F. Buonamici, *De Motu*, 1591, pp. 444–53.

10 Charles B. Schmitt, 'Filippo Fantoni, Galileo's Predecessor as Mathematics Lecturer at Pisa', *Studies in Renaissance Philosophy and Science X*, pp. 53–62.

11 Michael Segré, 'Viviani's Life of Galileo', *Isis* 80 (1989), pp. 207–31.

12 See Thomas B. Settle, 'Ostilio Ricci, a Bridge between Alberti and Galileo', *Actes du XIIe Congrès International d'Histoire des Sciences, Paris, 1968*, [published] 1971, tome 3B, pp. 121–6, where he gives evidence that Ricci used Alberti's *Ludi mathematici*. For a summary of Ricci's career see Arnaldo Masotti's article in *Dictionary of Scientific Biography*, vol. 11, pp. 405–6. For the importance of Galileo's training in perspective in making and depicting his first telescopic discoveries see Samuel Y. Edgerton, Jr., *The Heritage of Giotto's Geometry*, 1991, pp. 223–53 [with illustrations] and Martin Kemp, *The Science of Art: Optical Themes in Western Art from Brunelleschi to Seurat*, 1990, pp. 93–8 [with illustrations].

13 Adriano Carugo and Alistair C. Crombie, 'The Jesuits and Galileo's Ideas of Science and Nature', *Annali dell'Istituto e Museo di Storia della Scienza di Firenze*, Anno VIII, 1983, Fascicolo 2, pp. 12–13.

14 Galileo's undated tables of specific gravities of metals and jewels are in *EN* 1: 225–8 and his notes on Archimedes's *De Sphaera et Cylindro*: 233–42.

15 *EN* 19: 36, with the date 1587, so it could be as late as March 1588, according to when the unknown writer started the year.

16 See the edition of Jesuit educational documents by Ladislaus Lukács, *Monumenta paedagogica Societatis Jesu*, Nova editio penitus retractata, V, *Ratio atque Institutio Studiorum Societatis Jesu (1586, 1591, 1599)*, 1986, pp. 8*–9*. For Pereira's published views see his *De communi omnium rerum*

naturalium principiis & affectionibus libri quindecim, 1576, fol. 4v.
17 Ladislaus Lukács (ed.), *Monumenta paedagogica*, vol. V, p. 397. I am paraphrasing the 1599 *Ratio*, which does not differ on these points from Jesuit policy at the time when Galileo visited Clavius.
18 See Robert S. Westman, 'The Astronomer's Role in the Sixteenth Century: A Preliminary Study', *History of Science* 18 (1980), pp. 105–48.
19 Alistair Crombie pointed out to me the importance of Clavius's philosophy of science in conversation in 1974. For a fullish summary see Adriano Carugo and Alistair C. Crombie, 'The Jesuits and Galileo's Ideas of Science and of Nature', pp. 19–21.
20 Lukács, *Monumenta paedagogica*, vol. V, p. 402, gives the rules for the professor of mathematics in the 1599 *Ratio*. The 1586 *Ratio* (pp. 109–10 and 177) gives reasons for including mathematics in the syllabus and appoints Clavius to train Jesuit mathematicians who can serve elsewhere. See also Riccardo G. Villoslada, *Storia del Collegio Romano dal suo inizio (1551) alla soppressione della Compagnia di Gesù (1773)*, Analecta Gregoriana 66, 1954.
21 *EN* 10: Guidobaldo to Galileo: 25–6; 31; 33; 37; 39; Galileo to Guidobaldo 36–7; Moletti 21–2 and 1: 183.

CHAPTER 3 PROFESSOR AT PISA

1 Stillman Drake, 'Physics and Tradition before Galileo', in *Galileo Studies: Personality, Tradition and Revolution*, 1970, pp. 19–42; and *Galileo at Work: His Scientific Biography*, 1978, p. 20.
2 See the works by Drake in the preceding note. Galileo recognized that Philoponus and many Peripatetic writers had disagreed with Aristotle. He thought, however, that he could show what was wrong with Aristotle's account and provide a better alternative (*EN* 1: 284). See also Richard Sorabji (ed.), *Philoponus and the Rejection of Aristotelian Science*, 1987.
3 The experiment or demonstration is in Borro's *De motu gravium et levium*, 1576, p. 215. (I have not seen the first edition of 1575.) Galileo mentions Borro's thoroughness in his early dialogue on motion (*EN* 1: 367–8). There is an excellent study by Anna De Pace, 'Galileo lettore di Girolamo Borri nel *De Motu*', in *De Motu: Studi di storia del pensiero su Galileo, Hegel, Huygens e Gilbert*, Università degli Studi di Milano. Facoltà di Lettere e Filosofia, Quaderni di Acme 12, 1990.
4 I. Bernard Cohen, *The Birth of a New Physics*, 2nd edn, 1985, Supplement 3, pp. 194–5, with Thomas B. Settle's explanation of how differential muscular fatigue leads to the earlier release of the lighter object.
5 Pereira, *De communibus omnium rerum naturalium principiis & affectionibus*, 1576, p. 445.
6 Alistair Crombie and Adriano Carugo have shown, among other things, that Galileo's unpublished *Disputationes de praecognitionibus* bears unmistakable similarities to a work published by Ludovico Carbone in 1597; they do not accept William Wallace's conjecture that about 1590 Galileo used a

manuscript copy of the original lectures by the Jesuit Paolo della Valle, lectures which have not survived, though they were plagiarized by Carbone. So 1597, or thereabouts, is probably also the earliest date for other writings formerly classed as *juvenilia*. They also think it may be necessary to put the early writings on motion later than has been customary, though their main point is to establish that we lack any adequately grounded chronology for a significant part of Galileo's intellectual development. See especially their 'The Jesuits and Galileo's Ideas of Science and Nature', *Annali dell'Istituto e Museo di Storia della Scienza di Firenze*, Anno VIII, 1983, Fascicolo 2. It seems to me that the views on motion which I describe in this chapter are ones which Galileo must have worked out before, rather than after, his better-known investigations from 1602 onwards. They are, moreover, similar to views discussed by Mazzoni on pages 188–97 of his *In universam Platonis et Aristotelis philosophiam praeludia* of 1597, a work to which Galileo refers, expressing pleasure that Mazzoni seemed to have come round to views which Galileo held when they used to discuss matters in Pisa. See Frederick Purnell, 'Jacopo Mazzoni and Galileo', *Physis* 14 (1972), Fasc. 3, pp. 272–94. It is difficult to believe that on this point it was Galileo who was borrowing from Mazzoni, so it seems reasonable to treat these writings on motion as Pisan. This still leaves difficult questions of chronology about other writings.

7 See *Galileo Galilei: On Motion and Mechanics, Comprising 'De Motu'* (ca. 1590), translated with introduction and notes by I. E. Drabkin, and *Le Meccaniche* (ca. 1600), translated with introduction and notes by Stillman Drake, 1960.

8 This has not survived, but Carugo has discovered a mathematical treatise on cosmography, in a manuscript copy of 1606, which is an extensive commentary on the first book of the *Almagest*. See Carugo and Crombie, 'The Jesuits and Galileo's Ideas', p. 13. Galileo's comments on Aristotle and Archimedes are at *EN* 1: 292, 300, 302 and 307.

9 See Frederick A. Homann, 'Christopher Clavius and the Renaissance of Euclidean Geometry', *Archivum Historicum Societatis Iesu* 104 (1983), p. 238. Clavius is only listing opinions, but he seems to agree with one similar to that advocated by Galileo: see his *Euclidis Elementorum Libri XV*, 1574, fol. a7r.

10 Difference between medium and weight *EN* 1: 268–9; vacuum 276–84 and 294–6; Philoponus 284; falling bodies in theory and practice 272–3.

11 For samples of Benedetti's criticism of Aristotle in his *Diversarum speculationum mathematicarum et physicarum*, 1585, see Alexandre Koyré, *Études Galiléenes*, 1966, pp. 47–60.

12 S. Drake, 'Galileo's Steps to Full Copernicanism, and Back', *Studies in History and Philosophy of Science* 18 (1987), pp. 93–105, especially pp. 95–6, 98, 103 and 105.

13 See William A. Wallace, *Galileo's Early Notebooks: The Physical Questions. A Translation from the Latin, with Historical and Palaeographical Commentary*, 1977; *Galileo and His Sources: The Heritage of the Collegio Romano in Galileo's Science*, 1984. See also *Galileo Galilei, Tractatio de praecognitionibus et*

praecognitis and Tractatio de demonstratione, transcribed from the Latin auto-graph by William F. Edwards, with an introduction, notes and commen-tary by William A. Wallace, Antenore, Padova, 1988. Wallace has translated this: *Galileo's Logical Treatises: A Translation, with Notes and Com-mentary, of His Appropriated Latin Questions on Aristotle's Posterior Analytics*, Boston Studies in the Philosophy of Science 138, 1992. See also his compan-ion volume (137 in the same series). *Galileo's Logic of Discovery and Proof: The Background, Content, and Use of His Appropriated Treatises on Aristotle's Posterior Analytics.*

CHAPTER 4   THE PROPER HOME FOR HIS ABILITY

1 Ludovico Ariosto, *Orlando Furioso: (The Frenzy of Orlando)*, Translated with an introduction by Barbara Reynolds, Part One, 1977, Introduction, pp. 21–4.
2 See S. Drake, *Galileo at Work*, pp. 36–7.
3 See Winifred L. Wisan, 'The New Science of Motion: A Study of Galileo's *De motu locali*', *Archive for History of Exact Sciences* 13, 2/3 (1974), pp. 162–3.
4 See John Julius Norwich, *A History of Venice*, 1983, pp. 512–17.
5 An undated and cancelled fragment which attempted to prove the mis-taken principle is in *EN* 8: 383; the passage in the *Discourses* 8: 203–4.
6 Ronald H. Naylor, 'Galileo's Theory of Projectile Motion', *Isis* 71 (1980), pp. 550–70; quotation from Guidobaldo on p. 551.
7 Ronald. H. Naylor, 'Galileo's Method of Analysis and Synthesis', *Isis* 81 (1990), pp. 695–707, especially p. 697, from which I take this summary. Compare with Drake's 'Galileo's Experimental Confirmation of Horizon-tal Inertia', *Isis* 64 (1973), pp. 291–305.
8 Ronald H. Naylor, 'Galileo's Early Experiments on Projectile Trajectories', *Annals of Science* 40 (1983), pp. 391–6, especially pp. 391–2; see also Naylor's 'Galileo: The Search for the Parabolic Trajectory', *Annals of Science* 33 (1976), pp. 153–72.
9 See, for instance, David K. Hill, 'Dissecting Trajectories: Galileo's Early Experiments on Projectile Motion and the Law of Fall', *Isis* 79 (1988), pp. 646–68, and Stillman Drake and James MacLachlan, 'Galileo's Discovery of the Parabolic Trajectory', *Scientific American* 232, 3 (March 1975), pp. 102–10.
10 Alexandre Koyré, *Metaphysics and Measurement: Essays in the Scientific Revo-lution*, 1968.
11 On Tycho's observation of the new star and on parallax see Victor E. Thoren, *The Lord of Uraniborg*, 1990, pp. 55–70.
12 There is a fairly full discussion in S. Drake, *Galileo at Work*, pp. 104–10, partly revised in *Galileo: Pioneer Scientist*, pp. 130–2. See also his *Galileo Against the Philosophers in his Dialogue of Cecco di Ronchitti (1605) and Consid-erations of Alimberto Mauri (1606)*, in English translations with introduction

and notes by Stillman Drake, 1976.

13 Galileo's conjecture is described with bewilderment by Willy Hartner, 'Galileo's Contribution to Astronomy', in E. McMullin (ed.), *Galileo: Man of Science*, pp. 185–6.

14 Galileo's *Le operazioni del compasso geometrico et militare*, with dedication dated 10 July 1606, is *EN* 2: 365–424. Capra's *Usus et fabrica circini cuiusdam proportionis*, with dedication dated 7 March 1607, follows immediately. Galileo's *Difesa di Galileo Galilei*, published in Venice in 1607, is 2: 515–99.

CHAPTER 5 *DISCOVERIES AND CONTROVERSIES*

1 R. Westfall, 'Science and Patronage: Galileo and the Telescope', *Isis* 76 (1985), pp. 11–30, suggests that Galileo composed the anagram on the sole basis of his prior commitment to the Copernican system and only after being prompted by Castelli's query. Westfall consequently sees Galileo's procedure as a blemish on his reputation, but he does not support any attempt to play down Galileo's undoubted achievements. See Drake, *Galileo: Pioneer Scientist*, pp. 136–7.

2 See S. Drake, 'Galileo, Kepler and the Phases of Venus', *Journal for the History of Astronomy* 15 (1984), pp. 198–208, followed by Owen Gingerich's 'Phases of Venus in 1610', pp. 209–10, and William Peters's 'The Appearances of Venus and Mars in 1610', pp. 211–14.

3 *EN* 10: 484–5 Clavius to Galileo: note that no suspicion attaches to Clavius for not mentioning that the Jesuits were observing Venus; like Galileo they would not want to make a premature announcement; see pp. 499–502 for Galileo's reply.

4 Maelcote's lecture was printed from a manuscript copy by Favaro in *EN* 3: 291–8, under the title *Nuntius Sidereus Collegii Romani*. It was reported by Gregory of St Vincent, another oustanding Jesuit mathematician: *EN* 11: 163.

5 William R. Shea, *Galileo's Intellectual Revolution*, 1972, p. 23.

6 John North, 'Thomas Harriot and the First Telescopic Observations of Sunspots', in J. W. Shirley (ed.), *Thomas Harriot: Renaissance Scientist*, 1974, pp. 129–65, especially, pp. 132–4. Fabricius's book, *De maculis in sole observatis*, was published at Wittenberg.

7 Paul Guldin, *De centro gravitatis*, 1635, Prolegomena to Book I, p. 14.

8 I have used the account in Adam Tanner, *Universa theologia scholastica*, 1626, volume I, column 1726, paragraph 69. It does not differ significantly from what he wrote in his *Dissertatio peripatetica* of 1621.

9 See James Brodrick, *Robert Bellarmine: Saint and Scholar*, 1961, pp. 188–216.

10 For Aquaviva's letters see Richard J. Blackwell, *Galileo, Bellarmine, and the Bible: Including a Translation of Foscarini's 'Letter on the Motion of the Earth'*, 1991, especially pp. 137–42.

CHAPTER 6   *THE CONDEMNATION OF COPERNICANISM*

1   F. Buonamici, *De motu*, 1591, fol. 3a. The whole of the Galileo affair is fully documented in Favaro's edition of the works. See also Sergio M. Pagano (ed.), *I documenti del processo di Galileo Galilei*, 1984. An excellent work in English is *The Galileo Affair: A Documentary History*, edited and translated with an introduction and notes by Maurice A. Finocchiaro, 1989.

2   See Richard J. Blackwell, *Galileo, Bellarmine, and the Bible*, 1991; Foscarini's *Letter on the Motion of the Earth*, pp. 216–51.

3   Bellarmine's partial instrumentalism needs to be seen as part of a wider, partly anti-Aristotelian, interpretation of scripture in relation to astronomy: see *The Louvain Lectures (Lectiones Lovanienses) of Bellarmine and the Autograph Copy of his 1616 Declaration to Galileo*, texts in the original Latin (Italian) with English translation, introduction, commentary and notes by Ugo Baldini and George V. Coyne, Vatican Observatory Publications, *Studi Galileani*, vol. 1, no. 2, 1984.

4   Pereira (1535–1610) first published his *Commentariorum et Disputationum in Genesim* in 1591. I have used the revised edition, Cologne, 1601: the rule comes in Lib. 1, 8, para. 21; *EN* 5: 320.

5   Ibid. 9, para. 24.

6   Published in facsimile by the Scolar Press, 1970, as volume 22 of the English Recusant Literature Series: *A Most Learned and Pious Treatise . . . framing a Ladder, Wherby Our Mindes May Ascend to God, by the Stepps of his Creatures*, translated by T. B. Gent. [Francis Young], printed at Douay [actually secretly in England], 1616. The passage is on p. 45. The Latin original is *De ascensione mentis in Deum per scalas rerum creatarum opusculum*, Rome, 1615.

7   Christopher Hibbert, *The Rise and Fall of the House of Medici*, 1985, p. 221.

8   Augustine's *De Genesi ad Literam* can be found in Migne's *Patrologia Latina*, 34: this passage is vol. II, Cap. IX, para. 20, col. 270; *EN* 5: 318.

9   Ibid. vol. II, Cap. XVIII, para. 38, col. 280. *EN* 5: 310; 331–2.

10 Ibid. vol. II, Cap. IX, para. 20, col. 271.

11 Ernan McMullin, 'How Should Cosmology Relate to Theology?', in A. R. Peacocke (ed.), *The Sciences and Theology in the Twentieth Century*, 1981, especially, pp. 22–5.

12 *De Genesi ad Literam*, vol. I, Cap. XIX, para. 38, col. 260; *EN* 5: 339.

13 *The Sleepwalkers*, 1972 [1959], especially p. 503, [p. 495].

14 Jerome J. Langford, *Galileo, Science, and the Church*, rev. edn., 1971, p. 78.

15 Pope John Paul II, in a speech on the occasion of Einstein's centenary, 10 November 1979, *The Pope Teaches*, II, no. 4 (December 1979), p. 499. The original French text is in *Acta Apostolicae Sedis* 71 (1979), p. 1466.

CHAPTER 7   *CONTROVERSY AND NEW HOPE*

1   S. Drake, *Galileo at Work*, p. 278, referring to *EN* 6: 151. The Latin phrase is *illud testatum omnibus velim, nihil hic minus velle me quam pro Aristotelis*

placitis decertare. Earlier (p. 118) Grassi had said his contest was chiefly with Peripatetics.

2 *EN* 19: 400–01 Index's corrections; 13: 48–9 Barberini sends poem. The poem itself is in *Maphaei S. R. E. Card. Barberini nunc Urbani PP. VIII. Poemata*, 1634, pp. 278–82.

3 For a recent critical discussion see P. M. S. Hacker, *Appearance and Reality: A Philosophical Investigation into Perception and Perceptual Qualities*, (1987), 1991.

4 Urban's argument was reported by Cardinal Oreggi in his *De Deo Uno* of 1629: see, for instance, Libero Sosio's edition of the *Dialogue*, 1970, pp. 548–9.

5 Pietro Redondi, *Galileo eretico*, 1983. I have reviewed the translation by Raymond Rosenthal, *Galileo Heretic*, 1988 in *Studies in the History and Philosophy of Science* 21 (1990), pp. 685–90.

6 For a well-documented account which is very sympathetic to Riccardi and very critical of Galileo see Ambrogio Eszer, 'Niccolò Riccardi, O.P. – "padre Mostro"', *Angelicum* 60 (1983), pp. 428–61.

CHAPTER 8   *THE DIALOGUE AND GALILEO'S CONDEMNATION*

1 *Dialogo . . . Dove ne i congressi di quattro giornate si discorre sopra i due massimi sistemi del mondo tolemaico, e copernicano* etc. Text in *EN* 7: 25–530; English translation by Stillman Drake, *Dialogue Concerning the Two Chief World Systems – Ptolemaic & Copernican*, 2nd edn, 1970.

2 The idea can be found in earlier writers but most recent discussion has taken its origin from Thomas S. Kuhn's, *The Structure of Scientific Revolutions*, 1962, 2nd edn 1970. Kuhn's earlier *The Copernican Revolution: Planetary Astronomy in the Development of Western Thought*, (1957) 1959 is a very useful survey.

3 S. Drake, 'Sunspots, Sizzi, and Scheiner', in *Galileo Studies*, pp. 176–99.

4 C. Scheiner, *Rosa Ursina*, 1630, especially fols 18a, 30b and 42b.

5 For teaching at Douai in 1755 see a disputation poster in the library of Ushaw College, Durham. For useful background see my 'Copernicanism at Douai', *Durham University Journal* 67 (December 1974), pp. 141–8. On 16 April 1757 the Congregation of the Index decided to omit from future editions of the *Index of Prohibited Books* the general clause against all books teaching the immobility of the Sun and the mobility of the Earth (*EN* 19: 419). It took a separate incident in 1820 to lead to the omission of Galileo's *Dialogue* from the next edition of the *Index* (19: 420–1).

6 *Philosophiae naturalis principia mathematica; auctore Isaaco Newtono . . .* Perpetuis commentariis illustrata, communi studio PP. Thomae Le Seur & Francisci Jacquier, Editio altera . . . Geneva, 1760. The *declaratio* of these two priests, written in 1742, comes in the preliminary material of the third volume.

7   J. L. Russell, 'Catholic Astronomers and the Copernican System after the Condemnation of Galileo', *Annals of Science* 46 (1989), pp. 365–86.

CHAPTER 9   *TWO NEW SCIENCES*

1   Carugo's note is on p. 744 of his edition of the *Discourses* (see bibliography).
2   James MacLachlan, 'Galileo's Experiments with Pendulums: Real and Imaginary', *Annals of Science* 33 (1976), pp. 173–85, especially pp. 180–4. See also his 'A Test of an "Imaginary" Experiment of Galileo's', *Isis* 64 (1973), pp. 374–9.
3   Edoardo Benvenuto, *An Introduction to the History of Structural Mechanics. Part I: Statics and Resistance of Solids*, 1991, especially, pp. 145–97; p. 187 Guidobaldo. See also J. E. Gordon, *Structures: Or Why Things Don't Fall Down*, 1983.
4   *Reverendi Patris Dominici Soto Segobiensis . . . super octo libros physicorum Aristotelis quaestiones*, Secunda editio, Salamanca, 1551, fol. 96v.
5   T. B. Settle, 'An Experiment in the History of Science', *Science* 133 (1961), pp. 19–23.
6   R. H. Naylor, 'Galileo's Simple Pendulum', *Physis* 16 (1974), p. 24.
7   R. H. Naylor, 'Galileo and the Problem of Free Fall', *British Journal for the History of Science* 7 (1974), pp. 128, 131–3.
8   William Riley Parker, *Milton: A Biography*, 1968, especially, pp. 178–9 and 829–30.

CHAPTER 10   *REHABILITATION*

1   *Acta Synodalia Sacrosancti Concilii Oecumenici Vaticani II*, volumen III, pars VI, Vatican Polyglot Press, 1975, p. 268.
2   *Acta Synodalia Sacrosancti Concilii Oecumenici Vaticani II*, volumen IV, pars I, 1976, p. 544 ('clear allusion'); volumen IV, pars VII, 1978, pp. 755 and 758 (text of document and footnote). Translation in Austin Flannery (ed.), *Vatican Council II: The Conciliar and Post Conciliar Documents*, Dominican Publications, Dublin, 1975, p. 975 (paragraph 36 of the document).
3   The French text of the Pope's speech is in *Acta Apostolicae Sedis*, 71 (1979), pp. 1461–8.
4   The French text of the Pope's speech on 9 May 1983 is in *Acta Apostolicae Sedis*, 75 (1983), pp. 689–94.
5   The reference is to *Acta Apostolicae Sedis*, 71 (1979), p. 1541. Perhaps the clearest account of what the Pope has in mind is to be found in his message, published with the papers of a study week held at Castel Gandolfo in 1987 to celebrate the third centenary of the publication of Newton's *Principia*. See Robert J. Russell, William R. Stoeger and George V. Coyne (eds), *Physics, Philosophy, and Theology: A Common Quest for*

Understanding, 1988, separate pagination 1–14, after p. 14 of preface. See also the same editors' *John Paul II on Science and Religion: Reflections on the New View from Rome*, 1990.

6 I have used the text printed in the English weekly edition of *L'Osservatore Romano*, 4 November 1992, p. 8.

7 See the *Instruction on the Ecclesial Vocation of the Theologian*, published by the Congregation for the Doctrine of the Faith. English text in *L'Osservatore Romano* (weekly edition in English), 2 July 1990, pp. 1–4, especially paragraph 37; the Latin original is *Instructio de ecclesiali theologi vocatione* in *Acta Apostolicae Sedis* 82 (1990), pp. 1550–70, especially 1567.

8 *L'Osservatore Romano*, 4 November 1992, p. 1.

Bibliography

THE WORKS OF GALILEO GALILEI

Le Opere di Galileo Galileo, ed. Antonio Favaro, 20 vols, Florence, 1890–1909. This is the *Edizione Nazionale,* I have used the reprint (Florence: Barbèra) of 1968. I refer to it in the text and notes simply by volume and page number. To avoid confusion in the notes the abbreviation *EN* is sometimes used as a prefix.

Galileo's Early Notebooks: The Physical Questions. A translation from the Latin by William A. Wallace, with historical and palaeographical commentary. Notre Dame and London: University of Notre Dame Press, 1977.

Tractatio de praecognitionibus et praecognitis and Tractatio de demonstratione, Transcribed from the Latin autograph by William F. Edwards, with an introduction, notes and commentary by William A. Wallace. Padua: Antenore, 1988.

Galileo's Logical Treatises: A Translation, with Notes and Commentary, of His Appropriated Latin Questions on Aristotle's Posterior Analytics, by William A. Wallace. Boston Studies in the Philosophy of Science 138. Dordrecht, Boston and London: Kluwer, 1992.

On Motion and Mechanics, Comprising 'De Motu' (ca. 1590), translated with introduction and notes by I. E. Drabkin, and *Le Meccaniche (ca. 1600),* translated with introduction and notes by Stillman Drake. Madison: University of Wisconsin Press, 1960.

Galileo Against the Philosophers in his Dialogue of Cecco di Ronchitti (1605) and Considerations of Alimberto Mauri (1606), in English translations with introduction and notes by Stillman Drake. Los Angeles: Zeitlin & Ver Brugge, 1976.

Sidereus Nuncius or the Sidereal Messenger, translated with introduction, conclu-

sion, and notes by Albert Van Helden. Chicago and London: University of
Chicago Press, 1989.

*Cause, Experiment and Science: A Galilean Dialogue incorporating a new English
Translation of Galileo's 'Bodies That Stay atop Water, or Move in It'*, by Stillman
Drake. Chicago and London: University of Chicago Press, 1981.

Dialogue Concerning the Two Chief World Systems—Ptolemaic & Copernican, translated by Stillman Drake. 2nd edn. Berkeley and Los Angeles: University of
California Press, 1970.

Dialogo sopra is due massimi sistemi del mondo, tolemaico e copernicano, a cura di
Libero Sosio. Turin: Einaudi, 1970.

Dialogues Concerning Two New Sciences, translated by Henry Crew and Alfonso
de Salvio, with an introduction by Antonio Favaro. New York: Dover, 1954,
reprinting a 1914 edition.

Two New Sciences, Including Centres of Gravity & Forces of Percussion, translated,
with introduction and notes, by Stillman Drake. Madison: University of
Wisconsin Press, 1974.

Discorsi e dimostrazioni matematiche intorno a due nuove scienze, a cura di Adriano
Carugo e Ludovico Geymonat. Turin, Einaudi, 1958.

Discoveries and Opinions of Galileo, translated with an introduction and notes by
Stillman Drake. Garden City, New York: Doubleday, 1957.

Galilei, Galileo (and others), *The Controversy on the Comets of 1618*, Galileo
Galilei, Horatio Grassi, Mario Guiducci, Johann Kepler. Translated by
Stillman Drake and C. D. O'Malley. Philadelphia: University of Pennsylvania
Press, 1960.

OTHER REFERENCES

Ariosto, L., *Orlando Furioso: (The Frenzy of Orlando)*, translated with an introduction by Barbara Reynolds. Part One. London: Penguin Books, 1977.

Augustine, Saint, *De Genesi ad literam*, in J. P. Migne, *Patrologia Latina* 34 (1861).

Banfi, A., *Vita di Galileo Galilei*. 2nd edn. Milan: Feltrinelli, 1979.

Barbour, J. B., *Absolute or Relative Motion? A Study from a Machian point of view
of the Discovery and the Structure of Dynamical Theories*; Volume I: *The Discovery
of Dynamics*. Cambridge: Cambridge University Press, 1989.

Barker, P. and Ariew, R. (eds), *Revolution and Continuity: Essays in the History
and Philosophy of Early Modern Science*, Studies in Philosophy and the History
of Philosophy 24. Washington, DC: Catholic University of America Press,
1991.

Bellarmine, R., *A Most Learned and Pious Treatise . . . framing a Ladder, Wherby
Our Mindes May Ascend to God, by the Stepps of his Creatures*, translated by T.
B. Gent, [Francis Young], printed at Douay [actually secretly in England],
1616. Facsimile edition by Scolar Press, English Recusant Literature Series 22,
1970.

—— *The Louvain Lectures (Lectiones Lovanienses) of Bellarmine and the Autograph
Copy of his 1616 Declaration to Galileo*. Texts in the original Latin (Italian) with

English translation, introduction, commentary and notes by Ugo Baldini and George V. Coyne. Vatican Observatory Publications, *Studi Galileani* vol. 1, no. 2, 1984.

Benedetti, G. B., *Diversarum speculationum mathematicarum, & physicarum liber.* Turin, 1585.

Benvenuto, E., *An Introduction to the History of Structural Mechanics. Part I: Statics and Resistance of Solids.* New York and London: Springer, 1991.

Biagioli, M., 'Galileo's System of Patronage', *History of Science* 28 (1990), 1–62.

Blackwell, R. J., *Galileo, Bellarmine, and the Bible: Including a Translation of Foscarini's 'Letter on the Motion of the Earth'.* Notre Dame and London: University of Notre Dame Press, 1991.

Borro, G., *De motu gravium et levium.* Florence, 1576.

Brodrick, J., *Robert Bellarmine, Saint and Scholar.* London: Burns & Oates, 1961.

—— *Galileo: The Man, his Work, his Misfortunes.* London: Chapman, 1964.

Buonamici, F., *De motu libri X.* Florence, 1591.

Butts, R. E. and Pitt, J. C. (eds), *New Perspectives on Galileo.* University of Western Ontario Series in Philosophy of Science 14. Dordrecht and Boston: Reidel, 1978.

Cambridge History of Renaissance Philosophy, ed. Charles B. Schmitt. Cambridge: Cambridge University Press, 1988.

Campbell, I., 'Orders of Knowledge and Domains of Experience: Bellarmine and Galileo', *Atheism and Dialogue*, 19:4 (1984), 330–6 and 20:4 (1985), 377–92.

Carugo, A. and Crombie, A. C., 'The Jesuits and Galileo's Ideas of Science and Nature', *Annali dell'Istituto e Museo di Storia della Scienza di Firenze*, Anno VIII (1983), Fascicolo 2, 2–67.

Clavelin, M., *The Natural Philosophy of Galileo: Essay on the Origins and Formation of Classical Mechanics*, trans. A. J. Pomerans. Cambridge, Mass. and London: MIT Press, 1974.

Clavius, C., *Euclidis Elementorum Libri XV.* Rome, 1574.

Cohen, I. B., *The Newtonian Revolution, with Illustrations of the Transformation of Scientific Ideas.* Cambridge: Cambridge University Press, 1983.

—— *The Birth of a New Physics.* 2nd edn. New York and London: Norton, 1985.

—— *Revolution in Science.* Cambridge, Mass. and London: Harvard University Press, 1985.

Congregation for the Doctrine of the Faith, *Instruction on the Ecclesial Vocation of the Theologian. L'Osservatore Romano* (weekly edition in English), 2 July 1990, 1–4.

Crombie, A. C., *Augustine to Galileo.* Rev. edn. 2 vols. London: Penguin Books, 1969.

—— *Science, Optics and Music in Medieval and Early Modern Thought.* London and Ronceverte: Hambledon Press, 1990.

De Pace, A., 'Galileo lettore di Girolamo Borri nel *De Motu*', in *De Motu: Studi di storia del pensiero su Galileo, Hegel, Huygens e Gilbert*, Università degli Studi di Milano, Facoltà di Lettere e Filosofia, Quaderni di Acme 12. Milan: Cisalpino, 1990, pp. 3–69.

Dictionary of Scientific Biography, ed. C. G. Gillispie 16 vols. New York: Charles

Scribner's Sons, 1970–80.

Drake, S., *Galileo Studies: Personality, Tradition, and Evolution*. Ann Arbor: The University of Michigan Press, 1970.

—— 'Galileo's Experimental Confirmation of Horizontal Inertia', *Isis* 64 (1973), 291–305.

—— *Galileo at Work: His Scientific Biography*. Chicago and London: University of Chicago Press, 1978.

—— *Galileo*. Oxford: Oxford University Press, 1980.

—— 'Galileo, Kepler and the Phases of Venus', *Journal for the History of Astronomy* 15 (1984), 198–208.

—— 'Galileo's Steps to Full Copernicanism, and Back', *Studies in History and Philosophy of Science* 18 (1987), 93–105.

—— *Galileo: Pioneer Scientist*. Toronto, Buffalo and London: University of Toronto Press, 1990.

—— and J. MacLachlan, 'Galileo's Discovery of the Parabolic Trajectory', *Scientific American* 232:3 (March 1975), 102–10.

Duhem, P., *To Save the Phenomena: An Essay on the Idea of Physical Theory from Plato to Galileo*, trans. Edmund Doland and Chaninah Maschler. Chicago and London: University of Chicago Press, 1969.

Edgerton, S. Y. Jr., *The Heritage of Giotto's Geometry: Art and Science on the Eve of the Scientific Revolution*. Ithaca and London: Cornell University Press, 1991.

Eszer, A., 'Niccolo Riccardi, O.P. – "padre Mostro" ', *Angelicum* 60 (1983), 428–61.

Fabris, R., *Galileo Galilei e gli orientamenti esegetici del suo tempo*. Pontificiae Academiae Scientiarum Scripta Varia 62, Vatican City, 1986.

Favaro, A., *Galileo Galilei e lo studio di Padova*. 2 vols. Padua: Antenore, 1966, reprinting an 1883 edition.

Finocchiaro, M. A. (ed.), *The Galileo Affair: A Documentary History*, edited and translated with an introduction and notes. Berkeley, Los Angeles and London: University of California Press, 1989.

Fischer, K., *Galileo Galilei*. Munich: Beck, 1983.

Fölsing, A., *Galileo Galilei: Prozess ohne Ende. Eine Biographie*. Munich and Zurich: Piper, 1989.

Franklin, Allan, *The Principle of Inertia in the Middle Ages*, Boulder: Colorado Associated University Press, 1976.

Geymonat, L., *Galileo Galilei*. Turin: Einaudi, 1978.

Gingerich, O., 'Phases of Venus in 1610', *Journal for the History of Astronomy* 15 (1984), 209–10.

Gordon, J. E., *Structures: Or Why Things Don't Fall Down*, London: Penguin Books, 1983.

Grant, E., *Physical Science in the Middle Ages*. New York and London: John Wiley, 1971.

Guldin, P., *De centro gravitatis*. Vienna, 1635.

Hacker, P. M. S., *Appearance and Reality: A Philosophical Investigation into Perception and Perceptual Qualities*, Blackwell, Oxford (1987), 1991.

Hale, J. R., *Florence and the Medici: The Pattern of Control*, London: Thames and

Hudson, 1983.

Hall, A. R., *The Revolution in Science: 1500–1750*. London: Longman, 1983.

Harré, R. (ed.), *The Physical Sciences since Antiquity*. London: Croom Helm, 1986.

Hartner, W., 'Galileo's Contribution to Astronomy', in E. McMullin (ed.), *Galileo Man of Science*, 1967, pp. 185–6.

Hibbert, C., *The Rise and Fall of the House of Medici*. London: Allen Lane, 1974.

Hill, D. K. 'Dissecting Trajectories: Galileo's Early Experiments on Projectile Motion and the Law of Fall', *Isis* 79 (1988), 646–68.

Hintikka, J., Gruender, D. and Agazzi, E. (eds), *Theory Change, Ancient Axiomatics, and Galileo's Methodology*, Proceedings of the 1978 Pisa Conference on the History and Philosophy of Science, volume I. Dordrecht, Boston and London: Reidel, 1981.

Homann, F. A., 'Christopher Clavius and the Renaissance of Euclidean Geometry', *Archivum Historicum Societatis Iesu* 104 (1983), 233–46.

John Paul II, Speech on the Einstein Centenary. *Acta Apostolicae Sedis* 71 (1979), 1461–8.

—— Speech to the Gregorian University. *Acta Apostolicae Sedis* 71 (1979), 1538–49.

—— Speech of 9 May 1983, *Acta Apostolicae Sedis* 75 (1983), 689–94.

—— Message of 1 June 1988, in R. J. Russell and others, *Physics, Philosophy and Theology*, 1990, pp. M1–M14.

—— Speech of 31 October 1992 (rehabilitation of Galileo), *L'Osservatore Romano* (weekly edition in English), 4 November 1992, 1.

Kemp, M., *The Science of Art: Optical Themes in Western Art from Brunelleschi to Seurat*. New Haven and London: Yale University Press, 1990.

Kepler, J., *Kepler's Conversation with Galileo's Sidereal Messenger*. First complete translation, with an introduction and notes, by Edward Rosen, *The Sources of Science*, no. 5, New York and London: Johnson Reprint Corporation, 1965.

Koestler, A., *The Sleepwalkers: A History of Man's Changing Vision of the Universe*. London: Hutchinson, 1959; Penguin, 1972.

Koyré, A., *Études Galiléenes*. Paris: Hermann, 1966.

—— *Metaphysics and Measurement: Essays in the Scientific Revolution*. London: Chapman & Hall, 1968.

Kuhn, T. S., *The Copernican Revolution: Planetary Astronomy in the Development of Western Thought*. Cambridge, Mass.: Harvard University Press, 1957.

—— *The Structure of Scientific Revolutions*. 2nd edn. Chicago and London: University of Chicago Press, 1970.

Langford, J. J., *Galileo, Science, and the Church*. Rev. edn. Ann Arbor: University of Michigan Press, 1971.

Levere, T. H. and Shea, W. R. (eds), *Nature, Experiment, and the Sciences: Essays on Galileo and the History of Science in Honour of Stillman Drake*. Boston Studies in the Philosophy of Science 120. Dordrecht, Boston and London: Kluwer, 1990.

Lindberg, D. C. (ed.), *Science in the Middle Ages*. Chicago and London: University of Chicago Press, 1978.

—— *The Beginnings of Western Science: The European Scientific Tradition in Philo-*

sophical, Religious, and Institutional Context, 600 BC to AD 1450. Chicago and London: University of Chicago Press, 1992.

—— and Westman, R. S. (eds), *Reappraisals of the Scientific Revolution*. Cambridge: Cambridge University Press, 1990.

Lloyd, G. E. R., *Early Greek Science: Thales to Aristotle*. London: Chatto & Windus, 1970.

Lukács, L., *Monumenta paedagogica Societatis Jesu*, Nova editio penitus retractata, V, *Ratio atque Institutio Studiorum Societatis Jesu (1586, 1591, 1599)*. Rome: Institutum Historicum Societatis Jesu, 1986.

MacLachlan, J., 'A Test of an "Imaginary" Experiment of Galileo's', *Isis* 64 (1973), 374–9.

—— 'Galileo's Experiments with Pendulums: Real and Imaginary', *Annals of Science* 33 (1976), 173–85.

McMullin, E. (ed.), *Galileo: Man of Science*. New York and London: Basic Books, 1967.

—— 'How Should Cosmology Relate to Theology?' in A. R. Peacocke (ed.), *The Sciences and Theology in the Twentieth Century*, 1981, 17–57.

Mazzoni, J., *In universam Platonis, et Aristotelis philosophiam praeludia, sive de comparatione Platonis et Aristotelis*. Venice, 1597.

Naylor, R. H., 'Galileo and the Problem of Free Fall', *British Journal for the History of Science* 7 (1974), 105–34.

—— 'Galileo's Simple Pendulum', *Physis* 16 (1974), 23–46.

—— 'The Evolution of an Experiment: Guidobaldo del Monte and Galileo's *Discorsi* Demonstration of the Parabolic Trajectory', *Physis* 16 (1974), 323–46.

—— 'Galileo: The Search for the Parabolic Trajectory', *Annals of Science* 33 (1976), 153–72.

—— 'Galileo's Theory of Motion: Processes of Conceptual Change in the Period 1604–1610', *Annals of Science* 34 (1977), 365–92.

—— 'Galileo's Theory of Projectile Motion', *Isis* 71 (1980), 550–70.

—— 'Galileo's Early Experiments on Projectile Trajectories', *Annals of Science* 40 (1983), 391–6.

—— 'Galileo's Method of Analysis and Synthesis', *Isis* 81 (1990), 695–707.

North, J., 'Thomas Harriot and the First Telescopic Observations of Sunspots', in J. W. Shirley (ed.), *Thomas Harriot: Renaissance Scientist*, 1974, 129–65.

Norwich, J. J., *A History of Venice*. London: Allen Lane, 1982.

Pagano, S. M. (ed.), *I documenti del processo di Galileo Galilei*. Vatican City: Pontificia Academia Scientiarum, 1984.

Parker, W. R., *Milton: A Biography*. 2 vols. Oxford: Clarendon Press, 1968.

Peacocke, A. R., (ed.), *The Sciences and Theology in the Twentieth Century*. Stocksfield, Henley and London: Oriel Press, 1981.

Pedersen, O., 'Galileo and the Council of Trent: The Galileo Affair Revisited', *Journal for the History of Astronomy* 14 (1983), 1–29.

Pereira, B., *De communi omnium rerum naturalium principiis & affectionibus libri quindecim*. Rome, 1576.

—— *Commentariorum et disputationum in Genesim tomi quatuor*. Cologne, 1601.

Peters, W., 'The Appearances of Venus and Mars in 1610', *Journal for the History of Astronomy* 15 (1984), 211–14.

Pitt, J. C., *Galileo, Human Knowledge, and the Book of Nature: Method Replaces Metaphysics*. University of Western Ontario Series in Philosophy of Science 50. Dordrecht, Boston and London: Kluwer, 1992.

Poupard, Cardinal P. (ed.), *Galileo Galilei, 350 ans d'histoire*. Cultures et Dialogue 1. Paris: Desclée, 1983.

—— Speech of 31 October 1992, *L'Osservatore Romano* (weekly edition in English), 4 November 1992, 1.

Ptolemy, C., *Ptolemy's Almagest*, translated and annotated by G. J. Toomer. London: Duckworth, 1984.

Purnell, F., 'Jacopo Mazzoni and Galileo', *Physis* 14 (1972), Fasc. 3, 272–94.

Redondi, P., *Galileo eretico*, Turin: Einaudi, 1983. *Galileo Heretic*, trans, Raymond Rosenthal, London: Allen Lane, Penguin Press, 1988.

Righini Bonelli, M. L. and Shea, W. R. (eds), *Reason, Experiment and Mysticism in the Scientific Revolution*. London: Macmillan, 1975.

Roche, J., 'Harriot, Galileo, and Jupiter's Satellites', *Archives internationales d'histoire des sciences*, 32 (1982), 9–51.

Rose, P. L., 'The Origins of the Proportional Compass', *Physis* 10 (1968), 54–69.

Rosen, E., 'When Did Galileo Make His First Telescope?', *Centaurus* 1951, 44–51.

Rovasenda, E. di and Marini-Bettolo, G. B., *Federico Cesi nel quarto centenario della nascita*. Pontificae Academiae Scientiarum Scripta Varia 63, Vatican City, 1986.

Russell, J. L., 'Catholic Astronomers and the Copernican System after the Condemnation of Galileo', *Annals of Science* 46 (1989), 365–86.

Russell, R. J., Stoeger, W. R. and Coyne, G. V. (eds), *Physics, Philosophy, and Theology: A Common Quest for Understanding*. Vatican City: Vatican Observatory, 1988.

—— *John Paul II on Science and Religion: Reflections on the New View from Rome*. Vatican City: Vatican Observatory Publications, 1990.

Santillana, Giorgio de, *The Crime of Galileo*. London: Mercury Books, 1961.

Scheiner, C., *Rosa Ursina*. Bracciano, 1630.

Schmitt, C. B., *Studies in Renaissance Philosophy and Science*. London: Variorum Reprints, 1981.

—— *The Aristotelian Tradition and Renaissance Universities*. London: Variorum Reprints, 1984.

—— *Reappraisals in Renaissance Thought*, ed. Charles Webster. London: Variorum Reprints, 1989.

Segré, M., 'Viviani's Life of Galileo', *Isis* 80 (1989), 207–31.

Settle, T. B., 'An Experiment in the History of Science', *Science* 133 (1961), 19–23.

—— 'Galileo's Use of Experiment as a Tool of Investigation', in E. McMullin, *Galileo: Man of Science* 1967, pp. 315–37.

—— 'Ostilio Ricci, a Bridge between Alberti and Galileo', *Actes du XIIe Congrès International d'Histoire des Sciences, Paris, 1968*, [published] Paris, 1971, tome 3B, 121–6.

Sharratt, M., 'Copernicanism at Douai', *Durham University Journal* 67 (December 1974), 41–8.

—— 'Condemnation of Galileo', *Studies in the History and Philosophy of Science* 21 (1990), 685–90.

Shea, W. R., *Galileo's Intellectual Revolution*. London: Macmillan, 1972.

Shirley, J. W. (ed.), *Thomas Harriot: Renaissance Scientist*. Oxford: Clarendon Press, 1974.

—— *Thomas Harriot: A Biography*. Oxford: Clarendon Press, 1983.

Smith, L. P., *The Life and Letters of Sir Henry Wotton*. 2 vols. Oxford: Clarendon Press, 1907.

Sorabji, R. (ed.), *Philoponus and the Rejection of Aristotelian Science*. London: Duckworth, 1987.

Tanner, A., *Universa theologia scholastica*. 4 vols, Ingolstadt, 1626.

Thoren, V. E., *The Lord of Uraniborg: A Biography of Tycho Brahe*. Cambridge: Cambridge University Press, 1990.

Urban VIII, *Maphaei S.R.E. Card. Barberini nunc Urbani PP. VIII. Poemata*. Antwerp, 1634.

Van Helden, A., 'The Historical Problem of the Invention of the Telescope', *History of Science* 13 (1975), 251–63.

Villoslada, R. G., *Storia del Collegio Romano dal suo inizio (1551) alla soppressione della Compagnia di Gesù (1773)*. *Analecta Gregoriana* 66. Rome, 1954.

Viviani, V., *De Locis solidis secunda divinatio geometrica*, Florence, 1701 [1702].

Wallace, W. A., *Galileo and His Sources: The Heritage of the Collegio Romano in Galileo's Science*. Princeton, NJ: Princeton University Press, 1984.

—— *Galileo's Logic of Discovery and Proof: The Background, Content, and Use of His Appropriated Treatises on Aristotle's Posterior Analytics*. Boston Studies in the Philosophy of Science 137. Dordrecht, Boston and London: Kluwer, 1992.

Westfall, R., 'Science and Patronage: Galileo and the Telescope', *Isis* 76 (1985), 11–30.

Westman, R. S., 'The Astronomer's Role in the Sixteenth Century: A Preliminary Study', *History of Science* 18 (1980), 105–48.

Whitaker, E. A., 'Galileo's Lunar Observations', *Journal for the History of Astronomy* 9 (1978), 155–69 [with photographs].

Winkler, M. G. and Van Helden, A., 'Representing the Heavens: Galileo and Visual Astronomy', *Isis* 83 (1992), 199–217.

Wisan, W. L., 'The New Science of Motion: A Study of Galileo's *De motu locali*', *Archive for History of Exact Sciences* 13:2/3 (1974), 103–306.

Życiński, J. M., *The Idea of Unification in Galileo's Epistemology*, Studi Galileani, vol. 1, no. 4. Vatican City: Vatican Observatory Publications, 1988.

Index